Aurora

Guido Reni

SUN LORE
OF ALL AGES
A Collection of
Myths and Legends

WILLIAM TYLER OLCOTT

Lex Dei, Lux Diei

DOVER PUBLICATIONS, INC.
Mineola, New York

Bibliographical Note

This Dover edition, first published in 2005, is an unabridged republication
of the work originally published in 1914 by G. P. Putnam's Sons, New York and
London. The only alteration consists of moving some of the plates from their
original positions.

International Standard Book Number: 0-486-44556-9

Manufactured in the United States of America
Dover Publications, Inc., 31 East 2nd Street, Mineola, N.Y. 11501

To

MY MOTHER

Introduction

IN the compilation of the volume *Star Lore of All Ages*, a wealth of interesting material pertaining to the mythology and folk-lore of the sun and moon was discovered, which seemed worth collating in a separate volume.

Further research in the field of the solar myth revealed sufficient matter to warrant the publication of a volume devoted solely to the legends, traditions, and superstitions that all ages and nations have woven about the sun, especially in view of the fact that, to the author's knowledge, no such publication has yet appeared.

The literature of the subject is teeming with interest, linked as it is with the life-story of mankind from the cradle of the race to the present day, for the solar myth lies at the very foundation of all mythology, and as such must forever claim preeminence.

Naturally, there clusters about the sun a rich mine of folk-lore. The prominence of the orb of day, its importance in the maintenance and the development of life, the mystery that has ever

Introduction

enveloped it, its great influence in the well-being of mankind, have secured for the sun a history of interest equalled by none, to which every age and every race have contributed their pages.

In the light of modern science, this mass of myth and legend may seem childish and of trifling value, but each age spells its own advance, and the all-important present soon fades into the shadowy and forgotten past. It is therefore in reviving past history that progress is best measured and interpreted. The fancy so prevalent among the ancients that the sun entered the sea each night with a hissing noise seems to us utterly foolish and inane, but let us not ridicule past ages for their crude notions and quaint fancies, lest some of the cherished ideas of which we boast be transmuted by the touch of time into naught but idle visions.

It is therefore important for the student of history to study the past in all its phases, and whatever can be brought to light of the lore of bygone ages should have for us a charm and should find a place in our intellectual lives.

W. T. O.

Norwich, Conn.

Note

The thanks of the author are due The Macmillan Company, and Henry Holt & Company, for permission to use much valuable material from their copyrighted publications; and also to Professor A. V. W. Jackson, Mrs. J. R. Creelman, Mr. Leon Campbell, and the Smithsonian Institution, Bureau of American Ethnology, for rare illustrations.

Contents

Illustrations

Chapter I
Solar Creation Myths

Chapter I

Solar Creation Myths

IN the literature of celestial mythology, the legends that relate to the creation of the chief luminaries occupy no small part. It was natural that primitive man should at an early date speculate on the great problem of the creation of the visible universe, and especially in regard to the source whence sprang the Sun and the Moon.

This great question, of such vital interest to all nations since the dawn of history, presents a problem that is still unsolved even in this enlightened age, for, although the nebula hypothesis is fairly well established, there are astronomers of note to-day who do not altogether accept it.

The myths that relate to the creation of the sun generally regard that orb as manufactured and placed in motion by a primitive race, or by the God of Light, rather than as existing before the birth of the world. In other legends, the Sun was freed from a cave by a champion, or sprang into life as the sacrifice of the life of a god or hero.

3

These traditions doubtless arose from the fundamental belief that the Sun and the Moon were personified beings, and that at one time in the world's history man lived in a state of darkness or dim obscurity. The necessity for light would suggest the invention of it, and hence a variety of ingenious methods for procuring it found their way into the mythology of the ancient nations.

Of all the solar creation myths that have come down to us, those of the North American Indians are by far the most interesting because of the ingenuity of the legends, and their great variety. We would expect to find the same myth relating to the creation of the sun predominating, as regards its chief features, among most of the Indian tribes. On the contrary, the majority of the tribes had their own individual traditions as to how the sun came into existence. They agree, however, for the most part, in ascribing to the world a state of darkness or semi-darkness before the sun was manufactured, or found, and placed in the sky.

The great tribes of the North-west coast believe that the Raven, who was their supreme deity, found the sun one day quite accidentally, and, realising its value to man, placed it in the heavens where it has been ever since.

According to the Yuma Indian tradition, their great god Tuchaipa created the world and then

The Days of Creation
Burne-Jones

The Days of Creation
Burne-Jones

Permission of Frederick Hollyer

the moon.   Perceiving that its light was insufficient for man's needs, he made a larger and a brighter orb, the sun, which provided the requisite amount of light.

The Kootenays believed that the sun was created by the coyote, or chicken hawk, out of a ball of grease, but the Cherokee myth[1] that related to the creation of the sun was more elaborate, and seems to imply that the Deluge myth was known to them.

"When the earth was dry and the animals came down, it was still dark, so they got the sun and set it in a track to go every day across the island from east to west just overhead. It was too hot this way, and the Red Crawfish had his shell scorched a bright red so that his meat was spoiled, and the Cherokee do not eat it. The conjurers then put the sun another handbreadth higher in the air, but it was still too hot. They raised it another time, and another until it was seven handbreadths high, and just under the sky arch, then it was right and they left it so. Every day the sun goes along under this arch and returns at night on the upper side to the starting place."

This myth reveals a belief, common to many of the Indian tribes, that originally the sun was much nearer to the earth than now, and his scorching heat

[1] 19th Report U. S. Ethnology Bureau.

greatly oppressed mankind. Strangely enough, although it can be nothing but a coincidence, the nebular hypothesis of modern science predicates that the solar system resulted from the gradual contraction of a nebula. This implies that the planet earth and the sun were once in comparatively close proximity.

Among the Yokut Indians, there was a tradition that at one time the world was composed of rock, and there was no such thing as fire and light. The coyote, who of all the animals was chief in importance, told the wolf to go up into the mountains till he came to a great lake, where he would see a fire which he must seize and bring back. The wolf did as he was ordered, but it was not easy to take the fire, and so he obtained only a small part of it, which he brought back. Out of this the coyote made the moon, and then the sun, and put them in the sky where they have been to this day.

The significant feature of this myth is the fact, that, contrary to the general notion, the moon's creation antedated that of the sun. The explanation of this seeming incongruity appears in the legend of the Yuma Indians given above. The moon, although created first, did not give sufficient light, hence it was necessary to manufacture a source of light of greater power and luminosity.

The legends of the Mission Indians of California

reveal an altogether different view of the situation. In the myths cited above, the sun was manufactured to add to man's comfort. In the following legend the Earth-Mother had kept the sun in hiding, waiting for mankind to grow old enough to appreciate it; so, when the time came, she produced the sun and there was light.

In order that the Sun might light the world, the people of the earth decided that it must go from east to west, so they all lifted up their arms to the sky three times and cried out each time, "Cha, Cha, Cha!" and immediately the Sun rose from among them and went up to his appointed place in the sky.

One of the Mewan Indian sun myths reveals a novel tale to account for the presence of the sun. These Indians regarded the earth as an abode of darkness in primitive times, but far away in the east there was a light which emanated from the Sun-Woman.

The people wanted light very much, and appealed to Coyote-Man to procure it for them. Two men were sent to induce the Sun-Woman to return with them, but she refused the invitation. A large number of men were then sent to bring her back, even if they had to resort to force. They succeeded in binding the Sun-Woman, and brought her back with them to their land, where she ever after afforded people the light that is so necessary for their well-being. It was said that her entire body was covered with the beautiful iridescent shells of the abalone, and the light which shone from these was difficult to gaze upon.

According to a Wyandot Indian myth, the Little Turtle created the Sun by order of a great council of animals, and he made the Moon to be the Sun's wife. He also created the fixed stars, but the stars which "run about the sky" are supposed to be the children of the Sun and Moon.

The following Yuma legend[1] indicates that the moon was considered by their ancestors as of greater importance than the sun: "Kwikumat said, 'I will make the moon first.' He faced the east, and placing spittle on the forefinger of his right hand rubbed it like paint on the eastern sky until he made a round shiny place. 'I call it the moon,' said Kwikumat. Now another god, whom Kwikumat created, rubbed his fingers till they shone, and drawing the sky down to himself he painted a great face upon it rubbing it till it shone brightly. This he called the sun."

There is a tradition among the Pomo Indians of California, that, in very early times, the sun did not move daily across the heavens as it does now, but only rose a short distance above the eastern horizon each morning, and then sank back again.

[1] *Journal American Folk-Lore*, vol. xxii.

This arrangement did not suit people very well, and Coyote-Man determined to better conditions, so he started eastward to see what the trouble was with the sun.

He took with him some food, a magic sleep-producing tuft of feathers, and four mice. On the fourth day he arrived at the home of the sun people, who received him cordially, and a great dance was arranged in the dance house. In the midst of this building the sun was suspended from the rafters by ropes of grape-vines.

Coyote-Man liberated the mice, and told them to gnaw at the grape-vine ropes that held the sun. Meanwhile, he danced with the sun people, and, by the aid of the sleep-producing feathers, he succeeded in stupefying all the dancers. The mice by this time had freed the sun, which Coyote-Man seized and carried off with him. On his home-coming, the sun was laid on the ground, and the people discussed what should be done with it, but Coyote-Man decided that it should be hung up in the middle of the sky. This difficult task was delegated to the birds, but they all failed until it came the crows' turn. They were successful in hanging up the sun, and it has remained in its proper place ever since.

The Apache myth agrees with most of the Indian myths as regards the darkness that traditionally reigned over the world, and the peoples' desire for light, but their notion of the creation of the sun itself differs materially from the other Indian myths.

The myth relates that, originally, the only light in the world was that which emanated from the large eagle feathers that people carried about with them. This was such an unsatisfactory means of illuminating the world, that a council of the tribe was called for the purpose of devising a better system of lighting. It was suggested that they manufacture a sun, so they set about it. A great disk was made, and painted a bright yellow, and this was placed in the sky. The legend does not relate how this was accomplished. This first attempt at sun-making was not altogether successful, as the disk was too small. However, they permitted it to make one circuit of the heavens before it was taken down and enlarged. Four times it was taken down, and increased in size, before it was as large as the earth and gave sufficient light. Encouraged by their success at sun-making, the people made a moon, and hung it up in the sky. It appears that this light company's business, and its success, aroused the ire of a wizard and a witch who lived in the underworld. They regarded the manufacture of the sun and moon as presumptuous acts on the part of man, and attempted to destroy the luminaries; but the sun and moon fled from the underworld, leaving it in perpetual darkness, and found a safe abiding place in the heavens, where they have ever remained unmolested.

The Navajo Indian legend of the creation of the sun, moon, and stars is decidedly novel, and reveals the wide range of the imagination of primitive man in his conception of the creation of the celestial bodies.

The early Navajo, it appears, in common with many other Indian tribes, met in council to consider a means of introducing more light into the world; for, in early times, the people lived in semi-darkness, the obscurity resembling that of twilight.

The wise men concluded that they must have a sun, and moon, and a variety of stars placed above the earth, and the first work that was done to bring this about was the creation of the heavens in which to place the luminaries. The old men of the tribe made the sun in a house constructed for this special purpose, but the creation of the moon and stars they left to other tribes.

The work of creating the sun was soon accomplished, and the problem then presented itself of raising it up to the heavens and fixing it there. Two dumb fluters, who had gained considerable prominence in the tribe, were selected to bear the sun and moon (for they had also constructed a moon), Atlas-like, on their respective shoulders. These were great burdens to impose on them, for the orbs were exceedingly ponderous.   Therefore, the fluters staggered at first under the great weight that bore them down, and the one bearing the sun came near burning up the earth before he raised it sufficiently high; but the old men of the tribe lit their pipes and puffed smoke vigorously at the sun, and this caused it to retire to a greater distance in the heavens.

Four successive times they had to do this to prevent the burning up of the world, for the earth has increased greatly in size since ancient times; and, consequently, the sun had to be projected higher in the heavens, so that its great heat would not set the world on fire.

It will be noted that the language of the myths, as translated, has been followed closely.   This was in order to bring out more fully their quaint imagery.   The Indian was ever a poet, and a study of the tribal legends reveals many charming bits that would lose their beauty of expression were they transposed into the diction of our more prosaic tongue.

In the Ute Indian myth which follows, it appears that the sun had to be conquered and subjected to man's will, before it would perform its daily task in an orderly and regular way.

It is related that the Hare-God was once sitting by his camp-fire in the woods waiting the wayward Sun-God's return. Weary with watching, he fell asleep, and while he slumbered the Sun-God came, and, so near did he approach, that his great heat scorched the shoulders of the Hare-God. Realising that he had thus incurred the wrath of the Hare-God, the Sun-God fled to his cave in the underworld.

The Hare-God awoke in a rage, and started in pursuit of the Sun-God. After many adventures, he came to the brink of the world, and lay in wait for the object of his vengeance. When the Sun-God finally came out of his cave, the Hare-God shot an arrow at him, but the sun's heat burned it up before it reached its mark. However, the Hare-God had in his quiver a magic arrow which always hit the mark. This he launched from his bow, and the shaft struck the Sun-God full in the face, so that the sun was shattered into a thousand fragments. These fell to the earth and caused a great conflagration.

It was now the turn of the Hare-God to be dismayed at the results of his actions, and he fled before the destruction he had wrought. As he ran, the burning earth consumed all his members save his head, which went rolling over the face of the earth. Finally, it, too, became so hot that the eyes of the god burst, and out gushed a flood of tears which extinguished the fire.

But the Sun-God had been conquered, and awaited sentence. A great council was called, and, after much discussion, the Sun-God was condemned to pursue a definite path across the sky each day, and the days, nights, and seasons were arranged in an orderly fashion.

The following Cherokee Indian myth reveals the Sun as the arbiter of man's fate: "A number of people were engaged to construct a sun, which was the first planet made. Originally it was intended that man should live forever, but the sun, when he came to survey the situation, decided that, inasmuch as the earth was insufficient to support man, it would be better to have him succumb to death, and so it was decreed."

In *Creation Myths of Primitive America*, by Jeremiah Curtin, there is a particularly interesting solar myth. It is, therefore, given in much detail, as it is considered one of the most remarkable of the solar legends. As a pure product of the imagination, it ranks with the best examples of Egyptian and Grecian mythology.

The myth relates the efforts of a wicked and blood-thirsty old man named Sas,[1] to kill his son-in-law, Tulchuherris. After many ineffectual at-

[1] Sas was the Wintu Indian word for Sun.

tempts to accomplish his fell purpose, he proposed
a pine-bending contest, for he felt sure that, by
getting his son-in-law to climb to the top of a lofty
tree, he could bend it low, and, by letting go of it,
suddenly hurl the object of his enmity into the sky
and thus destroy him.

Tulchuherris had, however, a wise protector
hidden in his hair, in the guise of a little sprite
named Winishuyat, who warned him of his peril,
and enabled him to turn the tables on his wicked
father-in-law.

In the words of the myth: "He [Tulchuherris]
rose in the night, turned toward Sas, and said:
'Whu, whu, whu, I want you Sas to sleep soundly.'
Then he reached his right hand toward the west,
toward his great-grandmother's, and a stick came
into it. He carved and painted the stick beautifully, red and black, and made a fire-drill. Then
he reached his left hand toward the east, and wood
for a mokos [arrow straightener] came into it. He
made the mokos, and asked the fox dog for a fox-
skin. The fox gave it. Of this he made a head-
band, and painted it red. All these things he put
into his quiver. 'We are ready,' said Tulchuherris.
'Now, Daylight, I wish you to come right away.'
Daylight came. Sas rose, and soon after they
started for the tree. 'My son-in-law, I will go first,'
said Sas, and he climbed the tree. 'Go higher,'

said Tulchuherris, 'I will not give a great pull, go up higher.' He went high and Tulchuherris did not give a great pull so that Sas came down safely. Tulchuherris now climbed the tree, almost to the top. Sas looked at him, saw that he was near the top, and then drew the great pine almost to the earth, standing with his back to the top of the tree. Tulchuherris sprang off from the tree behind Sas, and ran away into the field. The tree sprang into the sky with a roar. 'You are killed now, my son-in-law,' said Sas, 'you will not trouble me hereafter.' He talked on to himself and was glad. 'What were you saying, father-in-law?' asked Tulchuherris, coming up from behind. Sas turned, 'Oh! my son-in-law, I was afraid that I had hurt you. I was sorry.' 'Now, my brother,' said Winishuyat, 'Sas will kill you unless you kill him. At midday he will kill you surely unless you succeed in killing him. Are you not as strong as Sas?'

" 'Father-in-law, try again, then I will go to the very top and beat you,' said Tulchuherris. That morning the elder daughter of Sas said to her sister after Sas had gone, 'My sister, our father has tried all people, and has conquered all of them so far, but to-day he will not conquer. To-day he will die. I know this. Do not look for him to-day. He will not come back. He will never come back to us.'

"Sas went up high. I will kill him now thought Tulchuherris, and he was very sorry, still he cried: 'Go a little higher. I went higher. I will go to the top next time. I will not hurt you, go a little higher.' Sas went higher and higher, till at last he said: 'I cannot climb any more, I am at the top, do not give a big pull, my son-in-law.' Tulchuherris took hold of the tree with one hand, pulled it as far as it would bend, pulled it till it touched the earth, and then let it fly. When the tree rushed toward the sky it made an awful noise, and soon after a crash was heard, a hundred times louder that any thunder. All living things heard it. The whole sky and earth shook. Olelbis, who lives in the highest place, heard it. All living things said: 'Tulchuherris is killing his father-in-law. Tulchuherris has split Sas.' The awful noise was the splitting of Sas. Tulchuherris stood waiting. He waited three hours perhaps, after the earth stopped trembling, then, far up in the sky he heard a voice saying: 'Oh, my son-in-law, I am split; I am dead. I thought I was the strongest power living, but I am not. From this time on I shall say Tulchuherris is the greatest power in the world.'

"Tulchuherris could not see any one. He only heard a voice far up in the sky saying: 'My son-in-law, I will ask you for a few things. Will you give

me your fox-skin head-band? Tulchuherris put
his hand into his fox-skin quiver, took out the
band, and tossed it to him. It went straight up
to Sas and he caught it. 'Now will you give me
your mokos?' Tulchuherris took out the mokos
and threw it. 'Give me your fire-drill.' He
threw that.

"Another voice was heard now, not so loud, 'I
wish you would give me a head-band of white
quartz.' This voice was the smaller part of Sas.
When Tulchuherris had given the head-band as
requested, he said: 'My father-in-law, you are
split. You are two. The larger part of you will
be Sas (the Sun), the smaller part Chanah (the
Moon), the white one, and this division is what you
have needed for a long time, but no one had the
strength to divide you. You are in a good state
now. You, Chanah, will grow old quickly and die,
then you will come to life, and be young again.
You will be always like that in this world. Sas,
you will travel west all the time, travel every
day without missing a day. You will travel day
after day without resting. You will see all things
in the world, as they live and die. My father-
in-law, take this too from me.' Tulchuherris then
threw up to Sas a quiver made of porcupine skin.
'I will take it,' said Sas, 'and I will carry it always.'
Then Tulchuherris gave Chanah the quartz head-

band, and said: 'Wear it around your head always, so that when you travel in the night you will be seen by all people.'

"Sas put the fox-skin around his head, and fastened the mokos crosswise in front of his fore-head. The fire-drill he fastened in his hair behind, placing it upright. At sunrise we see the hair of the fox-skin around the head of Sas before we see Sas himself. Next Tulchuherris threw up two red berries saying: 'Take these and make red cheeks on each side of your face, so that when you rise in the morning, you will be bright and make everything bright.'

"Tulchuherris then went west and got some white roots from the mountains, and threw them up to Sas saying: 'Put these across your forehead.' Next he stretched his right hand westward, and two large shells, blue inside, came to his palm. He threw these also up to Sas saying: 'Put these on your forehead for a sign when you come up in the morning. There is a place in the east which is all fire. When you reach that place, go in and warm yourself. Go to Olelpanti now. Olelbis, your father, lives there. He will tell you where to go.'

"Sas therefore went to Olelpanti where he found a wonderful and very big sweat-house. It was toward morning, and Olelbis was sleeping.

Presently he was startled by a noise and awoke, and saw some one near him, He knew at once who it was. Sas turned to him and said: 'My father, I am split. I thought myself the strongest person in the world, but I am not; Tulchuherris is the strongest.' 'Well my son, Sas,' said Olelbis, 'where do you wish to be? and how do you wish to live?' 'I am come to ask you,' replied Sas. 'Well,' answered Olelbis, 'you must travel all the time, and it is better that you go from east to west. If you go northward and travel southward, I don't think that would be well. If you go west and travel eastward I don't think that will be well either. If you go south and travel northward I don't think that will be right. I think that best which Tulchuherris told you. He told you to go east, and travel to the west. He said there is a hot place in the east that you must go into and get hot before you start every morning. I will show you the road from east to west. In a place right south of this is a very big tree, a tobacco tree, just half-way between east and west. When you come from the east, sit down in the shade of the tree, rest a few minutes, and go on. Never forget your porcupine quiver or other ornaments when you travel. . . . Go to the east. Go to the hot place every morning. There is always a fire in it. Take a white oak staff, thrust the end of it into the fire, till it is one glowing coal. When you travel west-ward carry this burning staff in your hand. In summer take a manzanita staff, put it in the fire, and burn the end. This staff will be red-hot all the day. Now you may go east and begin. You will travel all the time day by day without sleep-ing. All living things will see you with your blowing staff. You will see everything in the world, but you will be always alone. No one can ever keep you company or travel with you. I am your father, and you are my son, but I could not let you stay with me.' "

Among the Yana Indians of California, there is a myth similar, in some particulars, to the preced-ing legend. It accounts for the creation of the moon, but, as the sun figures prominently in the myth, it is given in full. It relates that once a youth named Pun Miaupa ran away from home after a quarrel with his father, and came to the house of his uncle.

He told his uncle that he desired to win for his wife Halai Auna, the Morning Star, the youngest and most beautiful daughter of Wakara, the Moon. His uncle tried to persuade him from making the attempt, knowing the danger to which his nephew would be exposed, for Wakara always killed his daughter's suitors; but finding the youth obdurate, he set out with him for Wakara's house. After a long journey they reached their destination, and now the uncle, who was a magician, knowing his nephew was in great peril, entered his nephew's heart.

Wakara received the youth cordially, and after a time placing him in the midst of a magic family circle, performed an incantation, and the group were transported magically to the house of Tuina, the Sun. Tuina was wont to slay men by giving them poisoned tobacco to smoke, but, although he presented five pipes to Pun Miaupa, he smoked them all with no ill effects, being protected from harm by his uncle, who all the time dwelt in his heart; and Halai Auna was glad that Tuina had been foiled in his attempt to slay her suitor.

Pun Miaupa's uncle now came forth from his nephew's heart, and through his powerful magic caused a great deluge to descend, and Tuina, and Wakara, and their families were all drowned save Halai Auna; but, so great was her grief over the loss of her family, that the magician took pity on her and restored them to life. He then entered his nephew's heart, and all returned to Wakara's house. Wakara was still bent on killing Pun Miaupa, and proposed a tree-bending contest similar to that in which Sas and Tulchuherris engaged, but he, too, shared the fate of the wicked Sas and was hurled into the sky where he remained.

Pun Miaupa laughed and said: "Now my father-in-law, you will never come here to live again, you will stay where you are now forever. You will become small, and then you will come to life and grow large. You will be that way always, growing old and becoming young again." Thus the moon changes in its phases from small to great as it pursues its heavenly way.

The following beautiful solar creation myth is from Japan.[1] It is entitled "The Way of the Gods" and contains a reference to a floating cloud in the midst of infinite space, before matter had taken any other form. This well describes the original nebula from which scientists aver the solar system was evolved. It is strange to find in the pages of mythology that record the creation of the celestial bodies allusions, that, in the light of modern science, savour more of fact than of fancy.

The myth is as follows: "When there was neither heaven, nor earth, nor sun, nor moon, nor anything that is, there existed in infinite space the Invisible Lord of the Middle Heaven, with him there were two other gods. They created between them a Floating Cloud in the midst of which was a liquid formless and lifeless mass from which the earth was evolved. After this were born in Heaven seven generations of gods, and the last and most

[1] *The Child's Guide to Mythology*, Helen A. Clarke.

perfect of these were Izanagi and Izanami. These were the parents of the world and all that is in it. After the creation of the world of living things Izanagi created the greatest of his children in this wise. Descending into a clear stream he bathed his left eye, and forth sprang Amaterasu, the great Sun-Goddess. Sparkling with light she rose from the waters as the sun rises in the East, and her brightness was wonderful, and shone through heaven and earth. Never was seen such radiant glory. Izanagi rejoiced greatly, and said: 'There is none like this miraculous child.' Taking a necklace of jewels he put it round her neck, and said: 'Rule thou over the Plain of High Heaven.' Thus Amaterasu became the source of all life and light, the glory of her shining has warmed and comforted all mankind, and she is worshipped by them unto this day.

"Then he bathed his right eye, and there appeared her brother, the Moon-God. Izanagi said: 'Thy beauty and radiance are next to the Sun in splendour; rule thou over the Dominion of Night.'"

In Norse mythology it is said that Odin arranged the periods of daylight and darkness, and the seasons. He placed the sun and moon in the heavens, and regulated their respective courses. Day and Night were considered mortal enemies. Light came from above, and darkness from beneath, and in the process of creation the moon preceded the sun.

There is, however, a Norse legend opposed to the view that the gods created the heavenly bodies. This avers that the sun and moon were formed from the sparks from the fire land of Muspelheim. The father of the two luminaries was Mundilfare, and he named his beautiful boy and girl, Maane (Moon) and Sol (Sun).

The gods, incensed at Mundilfare's presumption, took his children from him and placed them in the heavens, where they permitted Sol to drive the horses of the sun, and gave over the regulation of the moon's phases to Maane.

Among the Eskimos of Behring Strait the creation of the earth and all it contains is attributed to the Raven Father. It is related that he came from the sky after a great deluge, and made the dry ground. He also created human and animal life, but the rapacity of man threatened the extermination of animal life, and this so annoyed the Raven that he punished man by taking the sun out of the sky, and hiding it in a bag at his home. The people, it is said, were very much frightened, and disturbed at the loss of the sun, and offered rich gifts to the Raven to propitiate him; so the Raven relented somewhat, and would hold the sun up in one hand for a day or two at a time, so that the people could have sufficient light for hunting, and then he would put it back in the bag again.

This arrangement, though better than nothing, was not, on the whole, satisfactory to people, so the Raven's brother took pity on them, and thought of a scheme to better conditions. He feigned death, and after he had been buried and the mourners had gone away, he came forth from the grave, and took the form of a leaf, which floated on the surface of a stream. Presently, the Raven's wife came to the stream for a drink, and dipping up the water she swallowed with it the leaf. The Raven's wife soon after gave birth to a boy who cried continually for the sun, and his father, to silence him, often gave him the sun to play with. One day, when no one was about, the boy put on his raven mask and coat, and taking up the sun flew away with it, and placed it in its proper place in the sky. He also regulated its daily course, making day and night, so that ever thereafter the people had the constant light of the sun to guide them by day.

The ancient Peruvians believed that the god Viracocha rose out of Lake Titicaca, and made the sun, moon, and stars, and regulated their courses. Tylor[1] tells us that originally the Muyscas, who inhabited the high plains of Bogota, lived in a

[1] *Primitive Culture*, Edward B. Tylor.

state of savagery. There came to them from the east an old bearded man, Bochica (the Sun), who taught them agriculture and the worship of the gods. His wife Huythaca, however, was displeased at his attentions to mankind, and caused a great deluge which drowned most of the inhabitants of the earth.

This action angered Bochica, and he drove his wicked wife from the earth and made her the Moon (for heretofore there had been no moon). He then dried up the earth, and once more made it habitable, and comfortable for man to live in.

The Manicacas of Brazil regarded the sun as a hero, virgin born. Their wise men, who claimed the power of transmigration, said that they had visited the Sun, and that his figure was that of a man clothed in light; so dazzling was his appearance that he could not be seen by ordinary mortals.

According to Mexican tradition, Nexhequiriac was the creator of the world. He sent down the Sun-God and the Moon-God to illuminate the earth, so that men could see to perform their daily tasks. The Sun-God pursued his way regularly and unhindered, but the Moon-God, being hungry, and perceiving a rabbit in her path, pursued it. This took time, and then she tarried to eat it, but when she had finished her meal, she found her brother, the Sun, had outdistanced her, and was far ahead, so that ever thereafter she was unable to overtake him.   This is also the reason, says the legend, why the sun always appears to be ahead of the moon, and why the sun always looks fresh and red, and the moon sick and pale.   Those who gaze intently at the moon can still see the rabbit dangling from her mouth.

The Bushmen of South Africa have a curious legend to account for the creation of the sun.   According to this tradition, the Sun was originally a man living in a dwelling on earth, and he gave out only a limited amount of light, which was confined to a certain space about his house, and the rest of the country was in semi-darkness.   Strangely enough, the light which he gave out emanated from one of his arm-pits when his arm was upraised. When he lowered his arm, semi-darkness fell upon the earth.   In the day the Sun's light was white, but at night the little that could be seen was red like fire.   This obscure light was unsatisfactory to the people, and, just as we have seen the Indians of North America met in council to decide the question of a better light for the world, the Bushmen considered the matter, and an old woman proposed that the children seize the Sun while he was sleeping, and throw him up into the sky, so that the Bushman's rice might become dry for them, and the Sun make bright the whole world.

So the children acted on the suggestion, and approached the sleeping Sun warily; finally, they crept up close, and spoke thus to him: "Become thou the sun which is hot so that the Bushman's rice may dry for us, that thou mayst make the whole earth light, and the whole earth may become warm in summer." Therewith, they all seized the Sun by the arm, and hurled him up into the sky. The Sun straightway became round, and remained in the sky forever after to make the earth bright, and the Bushmen say that it takes away the moon and pierces it with a knife.

According to New Zealand traditions, the sun is the eye of Maui, which is placed in the sky, and the eyes of his two children became the morning and the evening star. The sun was born from the ocean and the story of its birth is thus related[1]: "There were five brothers all called Maui, and it was the youngest Maui who had been thrown into the sea by Taranga his mother, and rescued by his ancestor Great-Man-In-Heaven, who took him to his house and hung him in the roof. One night, when Taranga came home, she found little Maui with his brothers, and when she knew her last born, the child of her old age, she took him to sleep with her, as she had been wont to take his brothers before they were grown up. But the little Maui

[1] *Primitive Culture*, Edward B. Tylor.

became vexed and suspicious when he found that every morning his mother rose at dawn and disappeared from the house in a moment, not to return till nightfall. So one night he crept out, and stopped up every crevice in the wooden window, and the doorway, that the day might not shine into the house. Then broke the faint light of early dawn, and then the sun rose and mounted into the heavens, but Taranga slept on, for she knew not that it was broad daylight outside. At last she sprang up, pulled out the stopping of the chinks, and fled in dismay. Then Maui saw her plunge into a hole in the ground and disappear, and thus he found the deep cavern by which his mother went down below the earth as each night departed."

Another Maori legend relates that Maui once took fire in his hands, and when it burned him he sprang with it into the sea. When he sank in the waters the sun set for the first time, and darkness covered the earth. When Maui found that all was night he immediately pursued the sun and brought him back in the morning.

The Tonga tribe, of the South Pacific Islands, have a curious myth respecting the creation of the sun and moon. It appears that in primitive times, before there was any light upon the earth, Vatea and Tonga-iti quarrelled as to the parentage of a child. Each was confident the child was his, and to end the dispute they decided to share it. The infant was forthwith cut in two; Vatea took the upper half as his share, and squeezing it into a ball tossed it up into the sky where it became the sun. Tonga-iti allowed his share, the lower part of the infant, to remain on the ground for a day or two, but seeing the brightness of Vatea's half, he squeezed his share too, and threw it up into the dark sky when the sun was absent in the underworld, and it became the moon. Thus the sun and the moon were created, and the paleness of the moon is due to the fact that all the blood was drained out of it when it lay on the ground.

An Australian legend relates that the world was in darkness till one of their ancestors, who dwelt in the stars, took pity on them, and threw into the sky an emu's egg, which straightway became the sun.

The Dyaks have a myth that the sun and the moon were created by the Supreme Being out of a peculiar clay which is found in the earth, but is very rare and costly. Vessels made from this clay are considered holy and a protection against spirits. These people, and other tribes dwelling in mountainous regions where the sun's red disk disappears from sight each night behind the mountains, speak of the sun as setting in a deep cleft in the rocks.

There are numerous other myths, of many

lands, that tell in the language of imagination of
the birth of the sun and moon. Most of them,
however, are fundamentally identical with those
cited.

The close agreement in the traditions of many
of the primitive inhabitants of the earth, that there
was life in the world before the luminaries occupied
the heavens, and that people then lived in a state
of semi-darkness, is perhaps the most striking
feature of the myths.

In fact, these ancient legends reveal many
points that are of interest, especially where simi-
larities exist in the early traditions of widely
separated tribes. It seems extraordinary that
men, in different parts of the world, could have
independently conceived the grotesque notions
that often characterise the solar creation myths.

There seems to have been no limit to the fancies
concerning the creation of the heavenly bodies by
the ancients, but there is little doubt that, were
the world to-day deprived of the teachings of
science, our imaginative instinct would enable us
to construct a myth to account for the presence
of the sun, a myth that would be fully as fantastical
as any which mythology has bequeathed to us.

Chapter II
Ancient Ideas of the Sun and the Moon

Chapter II

Ancient Ideas of the Sun and the Moon

A STUDY of mythology reveals many legends and traditions that concern the Sun and the Moon as related in some way by family or marital ties. Consequently, in spite of the fact that the purpose of this volume is to afford a comprehensive review of solar lore, these legends were considered sufficiently interesting as a further revelation of the solar myth to be included.

In the early history of all people we find the Sun and Moon regarded as human beings, more or less closely connected with the daily life of mankind, and influencing in some mysterious way man's existence, and controlling his destiny. We find the luminaries alluded to as ancestors, heroes, and benefactors, who, after a life of usefulness on this earth, were transported to the heavens, where they continue to look down on, and, in a measure, rule over earthly affairs.

The chief nature of the influence supposedly

exerted by the Sun and Moon over men was parental.   In fact, the very basis of mythology lies in
the idolatrous worship of the solar great father,
and the lunar great mother, who were the first
objects of worship that the history of the race
records.

"The great father," says Faber,[1] "was Noah
viewed as a transmigratory reappearance of Adam,
and multiplying himself into three sons at the
beginning of each world, the great mother was the
earth considered as an enormous ship floating on
the vast abyss, and the ark considered as a smaller
earth sailing over the surface of the Deluge.   Of
these, the former was astronomically typified by
the sun, while the latter was symbolised by the
boat-like crescent of the moon."

As human beings, the Sun and Moon were naturally distinguished as to sex, although there is in
the early traditions concerning them no settled
opinion as to the sex assigned to each, nor their
relation to one another.   Thus, in Australia, the
Moon was considered to be a man, the Sun a woman,
who appears at dawn in a coat of red kangaroo
skins, the present of an admirer.   Shakespeare
speaks of the Moon as "she," while in Peru, the
moon was regarded as a mother who was both
sister and wife of the Sun, like Osiris and Isis in

[1] *The Origin of Pagan Idolatry*, G. S. Faber.

Egypt. Thus the sister marriages of the Incas had in their religion a meaning and a justification.

The Eskimos believed that the Moon was the younger brother of the female Sun, while the early inhabitants of the Malay peninsula regarded both the Sun and Moon as women. One of the Indian tribes of South America regarded the Moon as a man, and the Sun his wife. The story goes that she fell twice; the first time she was restored to heaven by an Indian, but the second time she set the forest blazing in a deluge of fire.

Tylor[1] tells us that the Ottawa Indians, in their story of Iosco, describe the Sun and Moon as brother and sister. "Two Indians," it is said, "sprang through a chasm in the sky, and found themselves in a pleasant moonlit land. There they saw the moon approaching as from behind a hill. They knew her at first sight. She was an aged woman with white face and pleasing air. Speaking kindly to them, she led them to her brother the sun, and he carried them with him in his course and sent them home with promises of happy life."

Other Indian tribes, such as the Iroquois, Athapascas, and Cherokees, regarded the Sun as feminine, and, as a whole, the North American myths represented the Sun and Moon more frequently

[1] *Primitive Culture*, Edward B. Tylor.

as brother and sister than as man and wife. In Central and South America, on the contrary, particularly in Mexico and Peru, the Sun and Moon were regarded as man and wife, and they were called, respectively, grandfather and grandmother.

This confusion in the sex, ascribed to the Sun and Moon by different nations, may have arisen from the fact that the day is mild and friendly, hence the Sun which rules the day would properly be considered feminine, while the Moon which rules the chill and stern night might appropriately be regarded as a man. On the contrary, in equatorial regions, the day is forbidding and burning, while the night is mild and pleasant. Applying these analogies, it appears that the sex of the Sun and Moon would, by some tribes, be the reverse of those ascribed to them by others, climatic conditions being responsible for the confusion.

In the German language, the genders of the Sun and Moon are respectively feminine and masculine, contrary to the rule of the Romance languages, whereas, in Latin, the Sun is masculine and the Moon feminine. In the Upper Palatinate of Bavaria to-day it is still common to hear the Sun spoken of as "Frau Sonne," and the Moon as "Herr Mond," and this story is told of them:

"The moon and sun were man and wife, but the moon proving too cold a lover, and too much addicted to sleep, his wife laid him a wager by virtue of which the right of shining by day should belong in future to whichever of them should be the first to awake. The moon laughed but accepted the wager, and awoke next day to find the sun had for two hours already been lighting up the world. As it was also a condition and consequence of this agreement that unless they awoke at the same time they should shine at different times the effect of the wager was a permanent separation, much to the affliction of the triumphant sun, who, still retaining a spouselike love for her husband, was, and always is, trying to repair the matrimonial breach."

From the conception that the Sun and Moon were husband and wife many legends concerning them were created, chief among these being the old Persian belief that the stars were the children of the Sun and Moon.

The primitive natives of the Malay peninsula believed that the firmament was solid. They imagined that the sky was a great pot held over the earth by a slender cord, and if this was ever broken the earth would be destroyed. They regarded the Sun and Moon as women, and the stars as the Moon's children. A legend relates that the Sun had as many children as the Moon, in ancient times, and fearing that mankind could not bear so much brightness and heat, the Sun and Moon agreed to devour their children.

The Moon pretended to thus dispose of hers, and hid them instead; but the Sun kept faith, and made way with all her children. When they were all devoured, the Moon brought hers out from their hiding-place. When the Sun saw them she was very angry, and pursued the Moon to kill her, and the chase is a perpetual one. Sometimes the Sun comes near enough to bite the Moon, and then men say there is an eclipse. The Sun still devours her children, the stars, at dawn, and the Moon hides hers during the daytime, when the Sun is near, only revealing them at night when her pursuer is far away. One of the native tribes of Northern India believes that the Sun cleft the moon in two for her deceit, and thus cloven and growing old again she remains.

The daughters of the Sun and Moon are represented in Finnish mythology as young and lovely maidens, seated sometimes on the border of a red shimmering cloud, sometimes upon a rainbow, sometimes at the edge of a leafy forest. They were surpassingly skilful in weaving, an accomplishment probably suggested by the resemblance borne by rays of light to the warp of a web. As might be expected in such a climate, the gods of the Sun, Moon, and stars are represented as serene and noble beings, dwelling in glorious palaces, possessing all that earth contained of beauty, and generally willing to share with mankind the knowledge of mundane affairs, which their penetrating rays and lofty position secured for them.

In Finnish, the appellation "Paeivae," came to mean Sun and Sun-God, "Knu," Moon and Moon-God. Paeivae had two sons, one of whom was named "Panu."[1]

There are many legends of the sun and moon that relate their disputes and marital troubles, for mythology reveals that as husband and wife the Sun and Moon did not live happily together.

In the Kanteletar (a collection of Finnish popular songs), an amusing tale is told which concerns the Sun, Moon, and Pole Star, who, as the story goes, were suitors for the hand of a beautiful maiden hatched from a goose egg. "The pole star was successful, and won the maiden, who objected to the moon, as there was nothing stable in his appearance, inasmuch as his face was sometimes narrow, and sometimes broad. Moreover, he had a bad habit of roving about all night, and remaining idle at home all day, which habit was highly detrimental to the true interest of a house-

---

[1] Herr Schiefner calls attention to the similarity between this name and the Sanscrit "bhâme," which he says is found in the Vedas, meaning the sun, and also an epithet for fire.

hold. She objected to the sun as he caused not only the heat in summer but also the cold in winter, and the variations of the weather. The pole star she accepted because he always came home punctually."

In Bavaria, a tale is told about a maid "who was drawn up by the moon, thereby incurring the sun's jealousy. The sun, to get even with her faithless consort, spying the girl's lover asleep in a wood took him up to herself. The maid and her lover perceiving themselves thus remote from one another were naturally anxious to meet again, and a great grief it was to the moon when he found that the maiden no longer cared for him. The tears he sheds in consequence are what we call 'shooting stars.'"

Thorpe[1] gives the following account of the relationship between the Sun and Moon:

"The moon and the sun are brother and sister. They are the children of Mundilföri, who, on account of their beauty, called his son 'Máni," and his daughter 'Sôl,' for which presumption the gods in their anger took the brother and sister, placed them in the sky, and appointed Sôl to drive the horses that draw the chariot of the sun, which the gods had formed to give light to the world. The names of the horses that bear her car

[1] *Northern Mythology*, Benjamin Thorpe.

are the 'Watchful,' and the 'Rapid,' and under their shoulders the gods placed an ice-cold breeze to cool them.

"A shield stands before the sun; if it were not for this, the great heat would set the waves and mountains on fire. Two fierce wolves accompany the sun, a widespread and popular superstition, one named 'Sköll' follows the sun, and which she fears will devour her, the other called 'Hati' runs before the sun, and tries to seize the moon,— and so in the end it will be. The mother of these wolves is a giantess, who dwells in a wood to the east of Midgard. She brought forth many sons who are giants and all in the form of wolves. One of this race called 'Managarm' is said to be the most powerful. He will be sated with the lives of all dying persons. He will swallow up the moon, and thereby besprinkle both heaven and air with blood, then will the sun lose its brightness, and the winds rage and howl in all directions. Mâni directs the course of the moon. He once took up two children from the earth as they were going from a well bearing a bucket on their shoulders. They followed Mâni, and may be observed from the earth."

Among the Indians of North America we find many legends relating to the Sun and Moon, who were regarded by them as living beings. They taught their children that the sun represents the eyes of the mighty Manito by day, the god the Indians worshipped, and that the moon and stars were his eyes by night, and that they could not hide their words or acts from him. We find practically this same belief among the natives of Australia, who regard the sun as the eye of the greater god, and the moon as the eye of the lesser god.

The following is Father Brébeuf's version of the Huron legend of "the white one and the dark one," an interesting bit of Indian mythology:

"The sun and moon were known to the Hurons as Iouskeha and Taoniskaron, respectively. When they were grown up they quarrelled and fought a duel. Iouskeha was armed with a stag horn, while Taoniskaron contented himself with some wild rose berries, persuading himself that as soon as he should thus smite his brother he would fall dead at his feet, but it happened otherwise, and Iouskeha struck him so heavy a blow in the side that the blood gushed forth in streams. Taoniskaron fled and from his blood which fell upon the land came the flints which the savages still call 'Taoniscara,' from the victims name. Iouskeha was regarded by the Indians as their benefactor, their kettle would not boil were it not for him, it was he who learned from the Tortoise the art of making fire. Without him they would have no luck hunting, and it is he who makes the corn grow.

"Iouskeha the sun takes care for the living, and all things concerning life, and therefore, says the missionary, they say he is good; but the moon, who is the creatress of earth and man, makes men die, and has charge of their departed souls, and they say she is evil. The sun and moon dwell together in their cabin at the end of the earth, and thither it was that four Indians made a mythic journey. The sun received the travellers kindly, and saved them from the harm the beautiful but hurtful moon must have done them."[1]

In early Iroquois legends, the Sun and Moon are god and goddess of day and night, respectively, and acquired the characters of the great friend and enemy of man, the good and evil deities.

The Cheyenne Indians have a legend that relates to a dispute that took place between the sun and moon as to which was superior. The Sun said that he was bright and glorious to behold, that he ruled the day, and that no being was superior to him. The Moon replied that he ruled the night, and looked after all things on earth, and kept all men and animals from danger and that he had no superior. The Sun retorted, "It is I who light up

[1] *Primitive Culture*, Edward B. Tylor.

the world. If I should rest from my work everything would be darkened, mankind could not do without me." Then the Moon replied: "I am great and powerful. I can take charge of both night and day, and guide all things in the world, It does not trouble me if you rest." Thus the Sun and the Moon disputed, and the day they spoke thus to each other became almost as long as two days, so much did they have to say to each other. Neither gained his point, although the Moon declared there were a great many wonderful and powerful beings on his side. He had reference to the stars.

The old theologists ascribed to both the Sun and Moon the guardianship of certain gates or doors in the firmament. These imaginary portals, they claimed, were in the two opposite tropics, and from them it was believed that all human souls were mysteriously born.

The Japanese believe that the moon is inhabited by a hare, and that the sun is the abode of a three-legged crow, hence the expression, "The golden crow, and the jewelled hare," meaning the sun and moon.

There was a belief current in ancient times, in countries remote from each other, that those great in authority, or of a superior order of society, were descendants of the Sun.

Among the Hindus, the members of the military caste to which the rajahs always belong are styled "Surya-bans," and "Chandra-bans," or Children of the Sun, and Children of the Moon.

"The first Egyptian dynasty is said to have been that of the 'Aurites,' or Children of the Sun, for the Oriental word 'Aur' denotes the solar light.

"The Persian sun-god Mithras bore the name of 'Azon-Nakis,' or the lord Sun, and from him both his descendants, the younger hero gods, and his ministers the magi, were denominated 'Zoni' and 'Azoni,' or the posterity of the Sun.

"Among the Greeks we find an eminent family distinguished by the name of the 'Heliadæ,' or Children of the Sun, and originally this family consisted of eight persons. The Greeks were also familiar with the Children of the Moon. This ancient title was the ancient appellation of the Arcadians.

"Among the early Britons it was acknowledged that the solar 'Hu' was the father of all mankind. The Incas of Peru traced their genealogy from the sun, and called themselves Children of the Sun."[1]

It is natural that the diurnal motion of the sun and moon should have stimulated the imagination of ancient man, and led him to account for this

[1] *The Origin of Pagan Idolatry*, G. S. Faber.

movement in a variety of fanciful ways. At an early date the idea was current that there was a subterranean world, into which the sun descended at nightfall, and traversed during the dark hours, to emerge from the cave in the east at dawn.

Among the red races, one tribe thought that the Sun, Moon, and stars were men and women, who went into the western sea every night, and swam out by the east. They say "the sun cometh forth every morning at the Place of Breaking Light." From this idea there rose the fancy that the Sun at evening was swallowed up by some monster, and the personified sun is a hero or a virgin who is swallowed and afterwards released or rejected, as in the Greek myths of Perseus and Andromeda, and Hercules and Hesione, the old Norse story of Eerick and the Dragon, and the more similar Teutonic myth of Little Red Riding Hood.

The Mexicans believed that "when the old sun was burnt out, and had left the world in darkness, a hero sprang into a huge fire, descended into the shades below, and arose deified and glorious in the east as 'Tonatiuh,' the sun. After him there leapt in another hero, but now the fire had grown thin, and he rose only in milder radiance as 'Metztli,' the moon." [1]

[1] *Primitive Culture*, Edward B. Tylor.

The Nimbus-Crowned Figure (Mithras)

Horses of the Sun

Robert Le Lorrain, Inprimerie Nationale, Paris

Photo by Giraudon

From persons possessing an all-powerful influence over mankind, the sun and moon came to be regarded as places where men could be consigned, and the extremes of heat and cold associated with them respectively gave rise to the idea that the sun and moon were places where men were sent in punishment for earthly misdeeds, there to suffer eternally for their sins.

The belief in the power of the sun and moon as persons to take up to them human beings from earth may next be shown to have had an influence over mythology. The sun and moon were both believed to possess this power of abducting mortals whom it pleased them to transport to the sky.

The Greeks of modern Epirus have a tale of a childless woman, who, having prayed to the Sun for a girl, gained her request, subject only to the condition that the girl be restored to the Sun when she became twelve years of age. When the child had reached that age, and was one day picking vegetables in the garden, whom should she meet but the Sun. That luminary bade her go and remind her mother of her promise. The mother, in terror and consternation, shut the doors and windows to keep her child safe from the sun, but unfortunately she forgot the keyhole, by which entrance the Sun penetrated and succeeded in carrying off his prey.

The Bushmen, almost the lowest tribe of South

Africa, have the same star lore and much the same mythology as the Greeks, Australians, Egyptians, and Eskimos. They believe that the Sun and Moon originally inhabited the earth and talked with men, but now they live in the sky and are silent.

The Homeric hymn to the Sun-God Helios, in the same way (as Professor Max Müller observes), regards the Sun as a half-god, almost a hero, who had once lived on earth. This mythological sojourn of the Sun and Moon on the earth seems to have stimulated the mind of primitive man, and has given rise to a wealth of legends and traditions which is to be found in the ancient history of all nations.

Chapter III

Solar Mythology

Chapter III

Solar Mythology

SOME one has said: "If no other knowledge deserves to be called useful but that which helps to enlarge our possessions or to raise our station in society, then mythology has no claim to usefulness. But if that which tends to make us happier and better can be called useful, a knowledge of mythology is useful, for it is the handmaid of literature, and literature is one of the best allies of virtue and promoters of happiness."

The solar myth, above all others, commands the attention and interest of the student of mythology, for it is the very basis of the science; it permeates the early history of all people, its influence has made itself felt in every age, and many of the customs that govern our lives to-day are of solar origin.

The sun, above all that human eyes behold, is the chief element in life, the very essence of our existence, and to its beneficent influences we owe all that we possess to-day, that is of worth. How few realise this fact. " 'Differentiated sun-shine,' is the striking and suggestive phrase used by John Fiske in his *Cosmic Philosophy* to stand for all things whatsoever to be found in this great world of ours; from the tiny sun-dew, hid in the secret abiding places of spreading swamp lands, and the inconsequent midget it opens its sticky little fist to grasp, to the great forest tree, and all-consequent man armed with his conquering broadaxe. It is merely a terse symbolic way of describing the processes of cosmic evolution from the sun as the original source and continuous guiding power of our own special universe."

"Back of the present sun figures in primitive and culture lore are the animistic conceptions of the sun such as that of Manabozho, or the great white hare, of Algonquin legend, or Indra the bull sun of India. In course of time the zoömorphic sun gives place to the anthropomorphic sun, and finally we arrive at such personifications of the sun as Osiris in Egypt, Apollo in Greece, and Balder in Norse mythology. Indeed it might almost be said that all the great steps in the onward march of the human race could be found recorded in the various and multiple personifications of the sun." [1]

Our ideas concerning natural phenomena are

[1] *Ancient Myths in Modern Poets*, Helen A. Clarke.

but the result of past ages of research in the fields of science; but when we come to a consideration of the phenomena that day and night present, in their ever-changing phases, we find it extremely difficult to clearly understand the mental viewpoint of primitive man regarding this continual change, for the uninterrupted sequence and constant repetition of this phenomena has dulled our faculties and it escapes our attention.

In ancient times, however, this continual daily process was closely observed and seriously considered, and the sun in all its aspects became at an early date in certain countries a personified godhead.

The expression "swallowed up by night" is now a mere metaphor, but the idea it conveys, that of the setting sun, was a matter of great importance to the ancients. However, the daily aspects of the sun were not alone matters of concern, the seasonable changes were closely observed, and the springtide sun, returning with youthful vigour after the long sleep in the night of winter, had a different name from the summer and autumnal sun. There are consequently, a multiplicity of names for the sun to be found in a study of primitive history and mythology, and an enormous mass of sun myths depicting the adventures of a primitive sun hero in terms of the varying aspects which the sun assumes during the day and year.

There was simply no limit to the images suggested by these aspects, as Sir George Cox puts it [1]:

"In the thought of these early ages the sun was the child of night or darkness, the dawn came before he was born, and died as he rose in the heavens. He strangled the serpents of the night, he went forth like a bridegroom out of his chamber, and like a giant to run his course. He had to do battle with clouds and storms, sometimes his light grew dim under their gloomy veil, and the children of men shuddered at the wrath of the hidden sun. His course might be brilliant and beneficent, or gloomy, sullen, and capricious. He might be a warrior, a friend, or a destroyer. The rays of the sun were changed into golden hair, into spears and lances, and robes of light."

From this play of the imagination the great fundamental solar myths sprang, and these furnished the theme for whole epics, and elaborate allegories. Out of this enduring thread there came to be woven the cloth of golden legend, and the wondrous tapestry of myth that illuminates the pages of man's history.

It is our pleasant task to review this variegated tapestry that fancy displays, and inspect the great treasure-house of tradition where the people of all

[1] *The Mythology of the Aryan Nations*, Sir George Cox.

ages have stored the richest gift of their imagination, the solar myth.

Perhaps the earliest sun myths are those founded on the phenomena of its rising and setting. The ancient dwellers by the seashore believed that at nightfall, when the sun disappeared in the sea, it was swallowed up by a monster. In the morning the monster disgorged its prey in the eastern sky. The story of Jonah is thought to be of solar origin, his adventure with the whale bearing a striking analogy to the daily mythical fate of the sun.

Goldhizer,[1] an eminent mythologist, claims that the Biblical story of Isaac is a sun myth, and the first Enoch, the son of Cain, is of pure solar significance. He is a famous builder of cities, a distinct solar feature, but the fact that he lived exactly 365 days, the length of the solar year, proclaims his solar character. Cain is a sun hero and among his descendants none but solar figures are to be found. Noah is clearly a mythical figure of the sun resting, the word Nôach denoting "him who rests."

The word which pre-eminently denotes the sun in the Semitic languages, the Hebrew "Shemesh," conveys the idea of rapid motion, or busy running about. Thus we see in the Psalms the sun likened to a giant or hero running a course.

Swift steeds were associated with the sun in

[1] *Mythology Among the Hebrews*, Ignaz Goldhizer.

Classical, Indian, Persian, and Hebrew mythologies, and in the Hebrew worship in Canaan, horses were dedicated to the sun, as indeed they were in Greece at a later date.

In the Veda the sun is frequently called "the runner," "the quick racer," or simply "the horse." This idea of the swift flight of the sun is further carried out by attributing wings to the sun, or dawn, and on the Egyptian and Assyrian monuments we find the winged solar disk inscribed.

From this it was but a step of the imagination to regard the sun as a bird, and when the sun set the ancients said: "The bird of day is weary, and has fallen into the sea." It is even thought that the hare is symbolic of Eastertide, for the very reason that fleetness of foot was its chief attribute. It is also a significant fact that the solar personification of the North American Indians was called "the Great White Hare."

"The more the Babylonian mythology is examined," says Sayce, "the more solar is its origin found to be, thus confirming the results arrived at in the Aryan and Semitic fields of research. With two exceptions only the great deities seem all to go back to the sun."

Of the mythology of Egypt, the eminent authority Renouf makes the statement: "Whatever may be the case in other mythologies, I look upon the sunrise and sunset, on the daily return of day and night, on the battle between light and darkness, on the whole solar drama in all its details that is acted every day, every month, every year, in heaven and in earth as the principal subject of Egyptian mythology."

The predominant mythological figures of Egypt were so much involved in the sun worship of that country, and to such an extent Sun-Gods, that a discussion of their personality and deeds pertains more properly to the chapter on Sun Worship, and is omitted therefore in this place.

There is one feature of solar mythology that is striking because of its universality, and that is the connection which the figures personifying the sun in various lands have with navigation. The Jewish Midrâsh compares the course of the sun to that of a ship, and curiously enough to a ship coming from Britain, which is rigged with 365 ropes (the number of days in the solar year), and to a ship coming from Alexandria which has 354 ropes (the number of days of the lunar year).

In Egypt we see on the monuments the figure of Ra, the Sun-God, in his boat sailing over the ocean of heaven. "The sun king Apollo is with the Greeks," says Goldhizer,[1] "the founder of navigation," and even the legendary Charon, the ferry-

[1] *Mythology Among the Hebrews*, Ignaz Goldhizer.

man of the underworld, is a development of the solar myth. The Roman Sun-God, Janus, is also brought into connection with navigation, and the Peruvian sun deity came to them from the sea, and took his leave of them in a ship which floated down a river to the sea where it vanished.

The ancient Egyptians called the sun "the Cat," for, "like the sun," says Horapollo, "the pupil of the cat's eye grows larger with the advance of day." The Egyptians imagined that a great cat stood behind the sun which was the pupil of the cat's eye.

The following sun myth found in India is quoted from *Anthropology* by Edward B. Tylor. It relates that: "Vâmana, the tiny Brahman, to humble the pride of King Bali, begs of him as much land as he can measure in three steps, but when the boon is granted, the little dwarf expands into the gigantic form of Vishnu, and striding with one step across the earth, another across the air, and a third across the sky, drives Bali into the infernal regions, where he still reigns. This most remarkable of all Tom Thumb stories seems really a myth of the sun, rising tiny above the horizon, then swelling into majestic power, and crossing the universe. For Vâmana the dwarf is one of the incarnations of Vishnu, and Vishnu was originally the sun. In the hymns of the Veda the idea of his three steps is to be found before it had become a story, when it was as yet only a poetic metaphor of the sun crossing the airy regions in his three strides."

The ancient Hindus enthroned the Sun-God in a burning chariot, and saw in his flashing rays spirited and fiery steeds arrayed in resplendent and gleaming trappings. Where we would say, "the sun is rising," or, "he is high in the heavens," they remarked, "the sun has yoked his steeds for his journey."

One of the common appellations for the sun in mythology is "the cow," and the sun's rays are described as the cow's milk. In the Veda this is one of the most familiar conceptions. These are good examples of the part imagination has played in the development of solar mythology. Given the notion that the sun is a chariot, the rays are seen immediately to resemble steeds, and, likewise, if the sun be likened to a cow, the rays must peradventure represent milk.

The sun's rays are compared more consistently with locks of hair or hair on the face or head of the sun. The Sun-God Helios is called by the Greeks "the yellow-haired," and long locks of hair and a flowing beard are mythological attributes of the sun in many lands. In an American Indian myth the Sun-God is described as an old man with a full beard, and the long beards of the Peruvian and Toltec Sun-Gods are often referred to in the mythological references concerning them.

If mythology is regarded as a wondrous piece of tapestry, wrought by imagination and fancy, displaying in many hues the noble deeds of gods and heroes of the ancient world, then, the part woven by the Greeks may well be considered the most conspicuous for brilliancy of conception and beauty of design of all that enters into this marvellous and priceless fabric.

It has been said that Greek mythology, in its dynastic series of ruling gods, shows an evolution from a worship of the forces of nature to a worship of the powers of the mind. It is beyond question the most complete in its details, the most perfect viewed from an artistic standpoint, the most beautiful and enduring of all the world's store of legendary lore that has come down to us, and in this wealth of mythology, the solar myth stands out supreme, as the central figure, clothed in the matchless imagery of a naturally poetical and highly artistic people.

In the following discussion of the Greek sun myths, there is much that seems so grotesque and fanciful as to border on absurdity, but the seriousness of the subject cannot be doubted, and, in order to understand it fully, with a true sense of appreciation, we must ever regard the legends as interpreting the natural phenomena of day and night. Bearing this fact in mind will enable us to grasp the significance of much that would otherwise be meaningless.

The daily motions and varying aspects of the living and energetic sun hero may be said to comprise the motif of almost every legend and myth bequeathed to us by the ancients.

As in the study of sun worship, the Sun-God Helios first occupies the scene as the central figure in a widely spread and popular cultus, we will first consider the legends that cluster about this mythical personage whom the Greek nation once revered and worshipped with all the fire of religious ardour.

The most interesting myth concerning Helios is that told of him in the *Odyssey*. It relates that when the hero Odysseus was returning to his home in Ithaca, the goddess told him of the verdant island of Trinacria, where the Sun-God Helios pastured his sacred herds, consisting of seven herds of cows and seven herds of lambs, fifty in each herd, a number which ever remained constant. Odysseus, desirous of visiting this fair isle, set out forthwith, having been warned by Circe to leave the herds of the Sun unmolested lest he suffer evil consequences. Having landed on the island, his companions enjoyed to the full the delightful climate, but as food was short, they ignored the warning of the goddess, slaughtered the Sun's best cattle, and feasted on them for six days, when they took their departure.

Helios, deeply incensed by their conduct, and grieving for his lost herds, in which he had taken great pride and pleasure, besought Jove's aid to punish them:

"O Father Jove, and all ye blessed gods
Who never die, Avenge the wrong I bear
Upon the comrades of Laertes' son
Ulysses, who have foully slain my beeves,
In which I took delight whene'er I rose
Into the starry heaven, and when again
I sank from heaven to earth. If for the wrong
They make not large amends, I shall go down
To Hades, there to shine among the dead.

The cloud-compelling Jupiter replied:
"Still shine, O sun: among the deathless gods
And mortal men, upon the nourishing earth,
Soon will I cleave, with a white thunderbolt
Their galley in the midst of the black sea."

And so it came about that through the might of Jove a frightful storm arose, which well-nigh wrecked the vessel of the wanton adventurers; but Odysseus, resorting to heroic measures, prevented the loss of the ship, and his companions thus escaped with their lives.

"Nothing is so common in Aryan Mythology," says Paley, "as the mention of cows or oxen in connection with the Sun. They seem to represent bright forms that appear to go forth in the form of luminous fleeting clouds from the home of the Sun in the east. The stealing, and recovering, or killing of these oxen is the subject of many tales in the early Greek legends."

> The isles of Greece, the isles of Greece:
> Where burning Sappho lived and sung,
> Where grew the arts of war and peace,—
> Where Delos rose, or Phœbus sprung:
> Eternal summer gilds them yet,
> But all except the sun, is set.
>
> BYRON.

We come now to the most conspicuous figure in Grecian mythology, the redoubtable hero, whose life-deeds furnish the theme for innumerable legends, songs, poems, epics, and many of the noblest and most beautiful conceptions in the world of art.

The fact that this distinguished and exalted personage personifies the sun, stamps Phœbus Apollo as the greatest and most widely known Sun-God that mythology and history have produced.

The story of his glorious birth at Delos, and his successful combat with the great serpent that Hera set in his path in the vale of Crissa, and the establishment of his oracle at Delphi, is related in the Chapter on Sun Worship as pertaining more particularly to his deification.

In the mythological biography of such a distinguished character as Phœbus Apollo we should expect, of course, to find allusions to his love affairs, and one of these is described in the myth of Daphne which follows:

Daphne, the personified Dawn, springs from the waters at the first flush of morning light, and as the beautiful tints of early day fade gradually in the light of the rising orb, Daphne flees from Apollo as he seeks to win her. In her flight she prays the gods to assist her, and it is related that, in answer to her prayer, she was transformed into a laurel tree, which was ever thereafter sacred to Apollo.

Another myth refers to a hunting trip in which Apollo was accompanied by his friend Hyacinthus. They engaged in a game of quoits, and Apollo cast a quoit which rebounded and struck Hyacinthus a fatal blow. Filled with remorse at the untimely death of his friend by his hand, he transformed Hyacinthus into a beautiful flower which ever after bore his name.

According to Murray[1] "the object of this myth was to point to the alternating decay and return of life in nature, which in this instance is conceived

[1] *Manual of Mythology*, Alexander S. Murray.

Apollo and Daphne

Bernini, Gallery Borghese, Rome

Apollo Belvedere

Museum of Vatican, Rome

under the form of a youth, the disc which was thrown by Apollo being clearly a symbol of the sun which scorches vegetation."

It is further related of Apollo that incensed at Zeus for causing the death of his son Æsculapius by a thunderbolt, he shot some of the Cyclopes, the forgers of thunderbolts. This brought down upon him the wrath of the Supreme Being, and Apollo was banished from Olympus. During his period of exile he served Admetus as a herdsman.

Keary[1] claims that Admetus is really one of the names for Hades, and this reference to his service under him indicates his descent into the underworld for the sake of purification. Here again we find the belief current that the sun at nightfall descends into the realms beneath the earth and waters.

During his term of banishment Apollo served Laomedon, the prince of Troy. As this master did not pay him the agreed amount Apollo brought down upon the city a dreadful pestilence which depopulated the entire neighbourhood. Vexed at his exile, Apollo joined with Poseidon in an effort to dethrone Zeus. The plot failed, and both gods were sentenced to assist in building the walls of Troy.

Apollo was far famed as a musician, and once

[1] *Outlines of Primitive Belief*, Charles F. Keary.

had a quarrel with Pan who claimed that the flute was a sweeter instrument than the lyre, which was Apollo's favourite instrument. They agreed to refer the matter to Midas, King of Lydia, who favoured Pan, and Apollo, in his displeasure at the verdict, punished Midas by causing his ears to lengthen till they resembled those of an ass. Apollo apparently brooked no rivalry in his musical accomplishments, for when Marsyas boasted that he excelled Apollo in flute playing, the latter had him flayed alive.

As a mighty warrior Apollo distinguished himself in the Trojan war when he took part against the Greeks. His fury was irresistible, and it is said whole ranks of fighting men fell as he charged into their midst.

Being possessed of eternal youth, and the most accomplished of athletes, Apollo came to be regarded as the patron of youthful athletic contests, and the Pythian games he instituted to commemorate his victory over the Python were celebrated in all lands.

Space does not permit of a complete recital of the many deeds of this famous sun hero. His favourite animals were the hawk and the swan; his tree, the bay. He was represented in the perfection of united manly strength and beauty. His long hair hangs loose, his brows are wreathed with bay, and in his hands he bears his bow and lyre.

The wonderful and famous Apollo Belvedere shows at the same time the conception which the ancients had of this benign deity, and the high degree of perfection to which they had attained in sculpture.

Few deities had more appellations than Phœbus Apollo. He was called Delian, Delphian, Clarian, etc., from the places of his worship. He was also referred to as "the Loxian God," from the ambiguity of many of his oracular predictions. Another appellation which the god bore was "Lycius" which means either the Wolf-God, or the Golden God of Light. He is also called "the Mouse-God," because he was regarded either as the protector or as the destroyer of mice. Other names for Apollo were "Silver-bowed," "Far Shooter," "Light Producer," "Well haired," "Gold-haired," "Gold-sworded."

"The likeness between Apollo and Achilles scarcely needs to be pointed out," says Keary.[1] "Each is the ideal youth, the representative of young Greece, that which was to become in after years Hellas."

In contrasting the character of the Sun-Gods, Helios and Apollo, we note a striking similarity.

[1] *Outlines of Primitive Belief*, C. F. Keary.

Both are conspicuous for their brilliant appearance, both possess powers of producing and destroying life, and weapons that are invulnerable. They are endowed alike with inexhaustible powers of creating happiness or sorrow, pleasure or torment, health or sickness. The exercise of these versatile faculties furnishes the theme for the major portion of that great mass of legends which constitute the essential elements of Aryan Mythology.

In the figure of Herakles we have a Sun-God and hero whose fame has gone afar into all lands, and every age since his time has likened its greatest deeds to the power and might he displayed in the accomplishment of his superhuman deeds.

Even to-day, the construction of the Panama Canal is often alluded to as a deed worthy of Hercules, and the adjective herculean has a firm place in the literature and phraseology of modern times.

The key to the sun myths that relate to the life and deeds of Hercules is found in the idea of the sun's subservience to nature's immutable laws. The sun has a daily task which it must perform. It has a path to travel from which it must not deviate. It has ever before it a life of toil from which it cannot swerve. "Nowhere," says Cox,[1] "is the unutterable toil and scanty reward of the sun brought out so prominently as in the whole

[1] *The Mythology of the Aryan Nations*, Sir George W. Cox.

legend, or rather in the mass of unconnected legends which is gathered round the person of Herakles."

Herakles was a son of Zeus and Alcmene. Through the malignant hate of Hera he was doomed from his birth to serve Eurystheus, and, not content by thus consigning him to a life of servitude, the goddess sent serpents to strangle him while he was but an infant. The hero, however, possessed godlike strength even in his tender years, and easily destroyed the serpents, much to the amazement of those in charge of him.

This struggle and triumph of the god over the serpents, is very like the successful combat Apollo waged against Python. It represents the great battle that the mythology of all lands presents, the subduing of the powers of darkness by the might of the omnipotent sun, the regent of light. This irresistible power, which is the chief attribute of the sun, is the predominant element in the character of the sun hero Herakles, and enables him to perform at the bidding of his master the twelve stupendous tasks that brought him endless and imperishable fame.

It is unnecessary to recite in detail these various labours. Every mythology does justice to the subject, and allusions to them appear in the arts and letters of every subsequent age. It remains, however, to point out the solar significance of these mighty deeds, and how, even in many details, they represent the sun's triumph over the obstacles that nature ever imposes.

The first labour was that of the conquest of the Nemean lion. The myth relates that, after slaying the ferocious beast, Herakles tore its skin off with his fingers, and thereafter it figured as his shield in many a fierce contest. The lion's skin has been likened to the "raiment of tawny cloud which the sun seems to trail behind him as he fights his way through the vapours whom he is said to overcome."

Herakles is next called upon to subdue the hideous Lernean hydra. This creature was possessed of many heads, one of which was immortal. The hero succeeded in this task by burning off the heads whenever they were raised to attack him. The immortal head he buried beneath a stone. As the beast was possessed of many heads, so the storm-wind must continually supply new clouds to vanquish the sun; but the lighter vapour and mist, the immortal head, is only conquered for a time. The sun easily burns up the heavy clouds, the mortal heads, but only hides temporarily the immortal head which rises again and again to daunt him. In this fight Herakles was attended by his friend Iolaus,—this name recalls that of

Iolê, signifying the violet-tinted clouds, the attendants of the sun in its serene moments.

"In the third month the sun enters the sign Libra, when the constellation of the Centaur rises, and in his third labour Herakles encountered and slew the Centaur. These comparisons are traceable throughout the year." For a detailed treatment of these myths the reader is referred to Anthon's Classical Dictionary.

In the madness of Herakles we see a further proof that he personifies the sun, for, as the sun rises, it increases in power until its heat destroys the fruits of the earth it loves. Herakles in his insanity kills his own children.

The marriage of Herakles with Hebe, the Goddess of youth, which took place after the performance of his twelve labours, denotes the renewal of the year at the end of each solar revolution.

It is in the last act of his life that Herakles best portrays his solar character. The poisoned coat presented by his wife Dejaneira is donned. It represents the clouds which rise from the waters and surround the sun like a dark raiment. Soon the poison infects the hero's system, inflicting pangs of anguish. Herakles tries in vain to cast it off, but the "fiery mists embrace him, and are mingled with the parting rays of the sun, and the dying hero is seen through the scattered clouds of the sky, tearing his own body to pieces till at last his bright form is consumed in a general conflagration."

In this death scene of the solar hero, and in the glories of his funeral pyre, we have the most famous sunset scene that has ever been presented for our contemplation. All the wondrous colouring that adorns the western sky at set of sun illuminates the canvas, and the reflection of the scene streams afar, lighting the waves of the Ægean with its clustering isles, and painting in enduring hues a scene that all nations proclaim the sublimest that nature offers to man's vision.

Anthon thus writes of the sun hero: "If Herakles be regarded as having actually existed, nothing can be more monstrous, nothing more at variance with every principle of chronology, nothing more replete with contradictions than the adventures of such an individual as poetry makes him to have been. But considered as the luminary that gives light to the world, as the god who impregnates all nature with his fertilising rays, every part of the legend teems with animation and beauty, and is marked by a pleasing and perfect harmony."

The Latin Hercules is indubitably identified with the Greek Herakles, and the legend of his life is identical with that of the Hellenic hero. It sets forth the same great struggle between the powers of light and darkness that we find in the primitive Hindu myth of Indra and Ahi, the source from which it sprang.

One of the solar legends which has come down to us in the simplest form is that of Sisyphus, who, so the myth relates, was condemned to spend his days in laboriously rolling a great stone to the summit of a hill. No sooner was the task accomplished than the stone rolled of its own volition to the base of the hill and his task began anew.

If we regard this as a sun myth, we see how closely the details of the legend apply to the daily course of the sun. It appears as a great sphere or ball which gradually mounts to the zenith each day as if laboriously propelled upward, on reaching the meridian it immediately begins its descent to the horizon.

Again, the sun by reason of its penetrating rays and its commanding position, suggests a power and a light from which nothing can be hid. The personification of this all-seeing eye would therefore be an all-wise being. "The Greek name Sisyphus," says Cox, "is simply a reduplicated form of Sophos, the wise, and so we have the image of a wise being compelled to ascend the heaven or mountain, and obliged in spite of his wisdom, his strength, and his power to come down as he had gone up. The idea of compulsion may soon pass into that of toil, and the latter into the thought of punishment, and thus the sun becomes a criminal under sentence."

In the myth of Ixîôn we have another solar legend. Ixîôn was condemned to a life of torture, being bound to a four-spoked and ever-revolving wheel. The name Ixîôn probably means "visitor," an appropriate name for the sun deity. The wheel revolves ceaselessly as the sun, and the condemned one is alternately raised into the high heavens, and lowered into the depths of the underworld. Cox[1] points out a curious but well-known characteristic of solar myths. "It is the identification of the sun both with the agent or patient, and with the thing or object by which the act is exercised. Ixîôn is the sun, and so is Ixîôn's wheel. Hercules is the sun who expires in the flames on the summit of Mount Œta, but the fiery robe which scorches him to death is the sun cloud."

The legend of Tantalus again reveals the fact that at one stage in the history of man, anthropomorphic ideas concerning the sun were prevalent. Tantalus was another victim of his misdeeds and consigned to eternal torture. It is related that he stood immersed in water to his chin, and yet dying of thirst, for as he lowered his head to drink, the water withdrew from him, and the earth appeared

[1] *The Mythology of the Aryan Nations*, Sir George W. Cox.

under his feet. To add to his torment luscious fruits hung alluringly from branches almost within his grasp, but no sooner did he stretch out his hands to pluck them than the wind blew them out of his reach, and their sight and that of the water only served to tantalise him.

The figure of the tortured Tantalus standing with his head alone exposed above the surface of the water clearly represents the sun setting in the western sea.

"The ancients speculated," says Paley, "on the hissing and steaming caused by the red hot orb being cooled down and extinguished in the sea. Fire and water could not co-exist, but in this myth the sun has the mastery, and it is the water that retires before the fire. Hence Homer says it was dried up by the god to punish Tantalus."

The word "Tantalus" means "the Poiser," the suspender in air of the huge disk of the sun, and one myth concerning Tantalus relates that, as a punishment for his evil deeds, he was suspended in mid air with a huge stone hanging over his head ever ready to fall and crush him, like a sword of Damocles. The ancients believed that the sun and earth were connected by a chain, and this fact reveals clearly the solar significance of the myth.

It is further related that Tantalus had had the honour of dining with Zeus. This signifies that the sun ascended from the mundane sphere to the upper regions.

Primitive man believed as the Bible teaches that all that exists will come to an end in a mighty conflagration, and that some day Tantalus himself would be hurled from his throne in the heavens and consume the earth.   Paley says: "If Sisyphus and Tantalus do not represent the Sun-Gods, the deeds and sufferings attributed to them have no intellectual point or meaning, the origin of such wild fables is quite incapable of explanation.   On the other hand, if they do, every detail in the narrative becomes simple and significant.   And if it can be shown even by a single example that the sun must be meant, then the doctrine of the solar myth is established."

Another beautiful myth of solar significance is that of Kephalos and Prokris of which the following is a brief version:

"Kephalos, a Phocian chief, coming to Athens won the love of Prokris, and plighted his faith to her.   But Kephalos was loved also by Eos, who sought to weaken his love for Prokris with a purpose so persistent that at last she induced him to make trial of her affection.   He therefore deserts Prokris to whom after a time he returns in disguise.   When in this shape he has won her love, he reveals himself, and Prokris in an agony of grief

Kephalos and Prokris

Guido Reni

Saint Michael

Guido Reni

and shame flies to Crete where she obtains from
Artemis the gift of a spear which shall never miss
its mark, and of a hound which can never fail to
seize its prey. With these gifts she returns to
Kephalos, who after seeing her success in the chase
longs to possess them. But they can be yielded
only in return for his love, and thus Prokris brings
home to him the wrong done to herself, and Eos
is for a time discomfited.

"But Prokris still fears the jealousy of Eos, and
watches Kephalos as he goes forth to hunt, until
one day while she lurked among the thick bushes,
she was fatally wounded accidentally by the un-
erring dart hurled by Kephalos.

"This myth explains itself. Kephalos is the
head of the Sun, and Kephalos loved Prokris, in
other words the Sun loves the dew, but Eos also
loves Kephalos, _i.e._, the dawn loves the Sun, and
thus at once we have the groundwork for her envy
of Prokris. So again when we are told that though
Prokris breaks her faith, yet her love is still given
to the same Kephalos different though he may
appear. We have here only a myth formed from
phrases, which told how the dew seems to reflect
many suns which are yet the same sun. The gifts
of Artemis are the rays which flash from each dew-
drop, and which Prokris is described as being
obliged to yield up to Kephalos, who slays her as

unwittingly as Phoibos causes the death of Daphne. The spot where she dies is a thicket in which the last dewdrops would linger before the approach of the mid-day heats."[1]

In the legend of Phaeton we have a more familiar sun myth. Phaeton, the son of Apollo, obtains his father's reluctant consent to drive for one day the chariot of the Sun. Hardly does he start upon his course, however, when the fiery steeds, realising that the reins are in inexperienced hands, run away, and the destruction of the world was threatened. Zeus hurled a thunderbolt at the unfortunate youth, and precipitated him into the river Eridanus.

"This myth reveals," says Cox, "the plague of drought, which made men say: 'Surely another, who cannot guide the horses, is driving the chariot of the sun.'"

The legend of Orpheus is akin to that of Daphne in its solar significance, as in this case Eurydice, although loved by the Sun, falls a victim to his radiancy, as he seeks to embrace her.

The myth of Meleagros reveals the capricious nature of the sun, its variations of light and shade being expressed by the alternate succession of swift deeds and moody fits of the hero when he retires sullenly from the sight of men.

---

[1] *The Mythology of the Aryan Nations*, Sir George W. Cox.

In the legend of Niobe the consuming power of the sun is manifested as aimed at those who have the temerity to face his dazzling brightness.

The tale of Althaia, regarded as a solar myth, relates to the destiny of the sun. In spite of its power and glory, it must die when the twilight hours usher in the night, but in the legend of Perseus we see once more in the slaying of the Medusa the victory of the sun over the powers of darkness. Perseus, the sun hero, is another Sun-God like Apollo, and Herakles. He has laborious tasks to perform for a tyrant master, invincible weapons, and a victorious career.

In the figure of Œdipus, the national hero of the Thebans, we have another Sun-God whose life and deeds were identical in many respects with those of the great sun heroes already referred to.

Orion, the mighty hunter, immortalised with Perseus and Herakles in the constellations, is also a personification of the sun, and in the splendour of his deeds exhibits the characteristics that made his predecessors famous.

The ancients attributed sudden deaths by sun-stroke to the shafts of the angered Sun-Gods, and the sun was thought to seek the sea each night for the purpose of bathing, so that thus purified it would rise to shine the next day with renewed lustre.

Paley tells us that there is a well-known legend,

the subject of the Homeric Hymn to Hermes, which is considered to have a solar significance. "That cunning god, the patron of rogues and thieves of every description, the Mercurius of the Romans, is said to have stolen and driven off a herd of cows while yet an infant. To prevent the theft being discovered by the traces of the animals, he fixed bundles of brushwood to their feet so that none could tell the direction they had taken. Now these cows are the clouds, the 'oxen of the sun' which figure so conspicuously in the *Odyssey*. It is a question of interest whether the Roman legend of the fire-breathing monster and robber Cacus, who stole the oxen of Hercules (the Sun-God), on his return from the west, is not in its origin identical. The story is told by Virgil, Propertius, and Ovid. It is said that one of the cows confined in the cave suddenly lowed, and led Hercules to the spot where he killed the robber and released the herd. The return of the lost sun after a thunder-storm explains the whole story very simply. The fire-breathing Cacus is the lightning, and the voice of the cow is the muttering of the thunder."

Space does not permit us to examine and discuss in detail the argument for the solar origin of the two great Homeric epics, the *Iliad* and the *Odyssey*. The prayer of Thetis to Zeus to do honour to her son, on which much of the action of the *Iliad* turns, is clearly the ascent of the Sun-God to heaven. "The might, the invincible prowess, the unwearied strength of the hero, and his powers of destruction and devastation, nay, even his divinely made shield, are merely attributes of the sun in his midday splendour."

The very fact that the sun myths have been so prolific is significant, and this is probably due to the fact that, as Fiske[1] points out, "the dramatic types to which they have given rise are of surpassing human interest." Thus they have endured through the ages, and in these myths and legends which adorn the rich pages of Grecian mythology we see man's effort to explain natural phenomena in human terms, to endow deity with man's heroic attributes, and to translate physical laws in the light of man's comprehension.

As for proofs that most of the Greek legends about the gods and heroes are of solar origin, it is pointed out that the same actions are attributed to them all. As Paley puts it: "They are all slayers of monsters or powerful foes; all court, or carry off, or return a bride; all grow up brave, all perform some wonderful feat, all go in quest of some lost treasure; generally they are exposed in infancy but survive to cause the death of their own parents. They perform set tasks or labours.

[1] *Myths and Myth Makers*, John Fiske.

They are faithless to their first loves, they are reunited to them in the end. The simple fact, as it appears to the sense, that the sun leaves the east, and yet is found there again on the very next day, was spoken of under the figure of a bridegroom torn from his bride, soon to be reunited. In the childhood of mankind the daily death of the sun was regarded as a reality. If he was born again it was not from any astronomical necessity so to say, but from the sufferance of nature or of Varuna, the sky god, or from his own benevolence to men, either of which might fail, and the casual eclipses and obscurations might become perpetual.

"The birth and death of the sun, his connection with the dawn, and his tremendous and victorious efforts to regain it were the one theme and topic of regard. He was talked about (though in a different sort of language) just as we are always talking and are never tired of talking of the weather. Hence it is that solar myths seem all in all."

Grecian mythology in its solar aspects is reflected in the legends and traditions of the Latins. Perhaps the most typical instance of this, and certainly the most familiar Roman myth that has come down to us of this nature, is portrayed in Guido Reni's beautiful fresco of *Aurora*. Properly it should have been designated "Apollo," for the central figure is that of the noble Sun-God, and he dominates the scene. "Surrounded by all the light tripping Hours, each a very queen of loveliness, Aurora the goddess of the dawn leads the throng."

The Romans actually believed that the sun was the wheel of Apollo's chariot. Each morning the god rose from the eastern sea, and drove his four spirited steeds across the sky, and in the evening he descended into the western sea. At night, he reposed in a golden boat which was borne along the northern edge of the earth to the rising point in the east.

"Antiquity," says the Abbé Banier,[1] "has transmitted to us the names of the four horses that drew the chariot of the sun. They were Erythous or the red, Acteon, the luminous, Lampos the resplendent, and Philogœus, the earth loving. The first denotes the sun rising, whose rays are then reddish. Acteon represents the time when the same rays shot through the atmosphere are more clear, that is about the ninth or tenth hour of the morning. Lampos figures noonday when this luminary is in all his strength and glory, and Philogœus represents the setting sun that seems to kiss the earth."

In the sun myths of all nations we find allusions to the capricious nature of the sun. Now it

[1] *The Mythology and Fables of the Ancients*, Abbé Banier.

smiles and gladdens the earth with its golden light, and, presently, displeased at man, shuns his presence, and hides sullenly for a time in gloomy solitude.

In a famous Japanese legend we have a description of the efforts of man to appease the Sun-Goddess when for a time she had absented herself from the sky.

The Sun-Goddess had taken refuge in a cave, and the earth knew not her light, and was dark and gloomy. The eight hundred of lesser deities took counsel as to the best means to propitiate the Goddess, and win once more her favour and her light. A great round copper mirror was procured to represent the sun's disk, and this was surrounded by a circle of saplings that indicated the rays of the radiant sun.

In the upper branches of the trees were hung balls representing the sacred jewel, and in the lower branches, blue and white pendants. A prayer was then recited by the chief priest who acted for the Emperor, and the service ended with a dance and the lighting of many fires. After a time, the Sun-Goddess yielded to the entreaties of man and left her gloomy cave for her heavenly throne, where her presence ever brings joy to the hearts of all mankind.

In another version of this myth the Sun-Goddess

is said to have waxed inquisitive at the noise of the singing and dancing at the entrance of her retreat, and ventured forth to see what was taking place. Beholding her beautiful self in the mirror, she stepped forth into the world once more, and "her glory filled the air with rosy radiance."

The propitiatory service is akin in many respects to the ceremonial, common among many primitive tribes, of producing sunshine. The first requisite of this rite was a mock sun, and the idea seems to have been that, by instituting an unusual ceremony, the curiosity of the hidden Sun would be aroused and she would come forth to see what was taking place.

Chapter IV

Solar Mythology (*Continued*)

Chapter IV

Solar Mythology (*Continued*)

IN Norse mythology we find, as we might expect, many solar myths and sun heroes, whose knightly qualities and redoubtable prowess enable them to accomplish seemingly impossible tasks, and vanquish the most formidable of foes.

Odin governs the high heavens, and the sun is referred to as Odin's eye. Thor rules in the clouds. He is identified as a Sun-God, and, like Hercules, distinguished himself as the enemy of the powers of cold and darkness. He conquers the frost giants. Heimdal's realm is the rainbow, and Balder rules the realm of light, but the sun affects them all. "It is," says Anderson,[1] "Odin's eye, Balder's countenance; Heimdal needs it for his rainbow, and still the sun itself rides as a beaming maid with her horses from morning until evening."

In the following graceful Finnish myth we find the sun represented as a lamp illuminating the halls of Vanna Issa, the Supreme Deity, and en-

[1] *Norse Mythology*, R. B. Anderson.

trusted by him to the care of two immortal servants, a youth and a maiden. "To the maiden who is called 'Evening Twilight,' the ancient Father saith: 'My daughter, unto thee I entrust the Sun, extinguish him and hide away the fire that no damage may ensue.' Then to the Dawn: 'My son, it is thy duty to rekindle the light for a new course. On no day is the light to be absent from the arch of heaven.'

"In winter he resteth a great while, but in summer his repose is short, and Evening Twilight gives up the dying light into the very hands of Dawn who straightway kindles it into new life. At such times they each take one look deep into the other's dark brown eyes; they press each other's hands, and their lips touch. Once a year only for the space of four weeks they come together at midnight. Then Evening Twilight layeth the dying light into the hands of Dawn, and a pressure of hands and a kiss make them happy, and the cheeks of Evening Twilight redden, and the rosy redness is mirrored in the sky till Dawn rekindles the light."

The following myth of the "Witch and the Sun's Sister" reveals another type of sun myth involving the actions of members of the Sun's family. In some of these myths the Sun's son figures, and however capricious the Sun may act, the legends that relate to his family indicate that they were kindly disposed toward humankind, and in many of the myths they act the part of benefactors.

In a country far remote there were once a king and a queen, who had a son named Prince Ivan, who was dumb from birth. A groom told the Prince that he was destined to have a sister who would prove to be a terrible witch, and advised him to flee lest harm befall him, He took the advice, and his father provided him with a swift horse on which he took his flight.

After wandering far he sought a dwelling-place in vain, first with two women, then with a giant who was uprooting trees, and lastly with a giant who was levelling mountains. Finally, he came to the dwelling of the Sun's sister, and she received him just as if he had been her own son. After a time, Prince Ivan longed to return to his old home, and persuaded the Sun's sister to allow him to depart. On his homeward journey he was enabled, by magic bestowed on him by the Sun's sister, to assist the two women and the giants who had refused to take him in because of trials which beset them.

On his home-coming his sister, the wicked witch, laid plans to destroy him; but he was warned in time by a mouse, and though closely pursued by the witch he escaped her clutches through the aid of those he had befriended and the Sun's sister.

The mythology of the North American Indians contains many examples of the solar myth. One of their Sun-Gods was Michabo, whose name signifies the Great Hare, the Great White One, or the God of the Dawn, and the East. It is said that he slept through the winter months, and in the fall when he was about to seek repose, he filled his great pipe, and the blue clouds of smoke that he exhaled drifted over the landscape filling the air with the veil-like haze of Indian Summer.

Michabo was regarded by the Indians as their common ancestor, and the ruler of the numerous tribes, the founder of their religious ritual and the inventor of their art of picture-writing. He controlled the weather, and was the creator and preserver of heaven and earth. The totem or clan which was dedicated to him was revered with the greatest respect.

In *Myths and Myth Makers*, by John Fiske, we read the following poetical description of Michabo, the Indian solar deity:

"From a grain of sand brought from the bottom of the primeval ocean he fashioned the habitable land, and set it floating on the waters till it grew to such a size that a strong young wolf running constantly died of old age ere he reached its limits. He was also like Nimrod a mighty hunter.[1] One of his footsteps measured eight leagues. The Great Lakes were the beaver dams he built, and when the cataracts impeded his progress, he tore them away with his hands. Sometimes he was said to dwell in the skies with his brother, the Snow, or like many great spirits to have built a wigwam in the far north in some floe of ice in the Arctic Ocean. In the oldest accounts of the missionaries he was alleged to reside toward the east. He is the personification of the solar life-giving power, which daily comes forth from its home in the east making the earth to rejoice."

A Modoc Indian myth relates that every day the Sun is utterly destroyed, and reduced to a heap of ashes, but inasmuch as the Sun is immortal, the disk lies dormant in the ashes waiting its summons to renewed life. Some one must, therefore, rouse the slumbering Sun each morning, as a slave is called to daily labour, and this office is performed by the morning star. At the summons to awake, the golden disk springs from the ashes rejuvenated, and goes forth to run his course. Here we have

---

[1] In this connection it is interesting to note the unexplained association of Orion, the personification of Nimrod, the mighty hunter, and the timid hare. In the constellations we find Orion and Lepus contiguous. That they were designed to be thus closely associated in the heavens cannot be doubted.

a legend similar to that of the Phœnix, the mythical bird which rose from lifeless ashes once in five centuries.

The following Cherokee legend is one of the most interesting solar myths related by the Indians:

"The Sun lived on the other side of the sky vault but her daughter lived in the middle of the sky directly above the earth, and every day as the Sun was climbing along the sky arch to the west, she used to stop at her daughter's house for dinner.

"Now the Sun hated the people of the earth because they could never look straight at her without screwing up their faces. She said to her brother, the Moon: 'My grandchildren are ugly, they grin all over their faces when they look at me.' But the Moon said: 'I like my younger brothers.' They always smiled pleasantly at him when they saw him in the sky for his rays were milder. The Sun was jealous and planned to kill all the people. So every day when she got near her daughter's house she sent down such sultry rays that there was a great fever, and the people died by hundreds. They went for help to the Little Men who said the only way to save themselves was to kill the Sun. The Little Men made medicine, and changing two men to snakes sent them to bite the old Sun when she came next day, but the light of the Sun blinded them, and they were unable to harm the Sun.

Again the Little Men were appealed to, and changing a man into a rattlesnake they sent him to bite the Sun, but instead of the Sun the snake bit the Sun's daughter, and she died from the bite. Now was the Sun sad and people did not die any more, but now the world was dark all the time because the Sun would not come out. Again they appealed to the Little Men who told them they must appease the Sun by bringing back her daughter from the ghost country. Seven men were chosen to seek the daughter, and bring her back in a box. They were charged not to open the box after she was put into it. They succeeded in their quest, and started home with the daughter safe in a box. She pleaded so hard to be let out that when they were almost home they opened the box only a little way, but this was enough, and something flew past them into a thicket, and they heard a red bird cry, 'Kwish,' 'Kwish,' in the bushes. They shut down the lid but when they got home the box was empty. The Sun had been glad when they started for the ghost country, but when they came back without her daughter she grieved and cried and wept until her tears made a flood upon the earth, and people were afraid the world would be drowned.

"They held another council and sent their handsomest young men and women to amuse her so that she would stop crying. They danced before the Sun and sang their best songs, but for a long time she kept her face covered and paid no attention until at last the drummer suddenly changed the song, when she lifted up her face and was so pleased at the sight that she forgot her grief and smiled."[1]

It is a significant fact that we find here a legend in respect to the propitiation of the sun identical with the Japanese myth related above, where, because of the retirement of the Sun-Goddess into her cave, men made every effort to conciliate her, and, finally, by a ceremony of singing and dancing, they won her back to her place in the sky.

Another Cherokee Indian myth relates that several young warriors once set out on a journey to the sunrise land. On the way they had many strange adventures, and finally they came to the sun's rising place, where the sky touches the ground. They discovered that the sky was an arch of solid rock hanging above the earth, and it seemed to swing slowly up and down, so that as it rocked it left a little opening at its base through which the sun rose each morning.

The adventurers waited for the sun to come out and presently it appeared. It had a human figure, but it was too bright to permit of their seeing its

[1] From the 19th Report of the Bureau of Ethnology.

features clearly. As soon as it had emerged through the opening they tried to leap through the narrow orifice before it was closed, but just as the first warrior was passing through, the rock rim of the sky closed and crushed out his life. The others were afraid to make the attempt after this fatality, and returned home, and the return trip took them such a long time that when they at length reached the end of their journey they were all old men.

In another version of this myth, three brothers undertook the journey, and the two younger ones succeeded in leaping through the opening. The older brother attempted to follow, but he was crushed by the great rock rim of the sky. The two successful brothers continued journeying in a land where everything is different, and presently met their elder brother. They all proceeded to the house of the Supreme Deity, whose messenger the Sun was, and were purified and built over. They now possessed magic qualities which enabled them to perform wonderful feats of speed and strength. After a time they returned to their native village, but, like Rip Van Winkle, no one knew them save an old woman.

One of the most beautiful of the solar myths of the Indians is the Algonquin "Legend of the Red Swan" which is as follows:[1]

[1] *Primitive Culture*, Edward B. Tylor.

"The hunter Ojibwa had just killed a bear, and begun to skin him when suddenly something red tinged all the air around. Reaching the shore of a lake the Indian saw it was a beautiful red swan whose plumage glittered in the sun. In vain the hunter shot his shafts, for the bird floated unharmed and unheeding, but at last he remembered three magic arrows at home which had been his father's. The first and second arrows flew near and nearer, the third struck the swan, and flapping its wings it flew slowly towards the sinking of the sun."

Longfellow has adapted this beautiful episode as a sunset picture in one of his Indian poems:

> Can it be the sun descending
> O'er the level plain of water?
> Or the Red Swan floating, flying,
> Wounded by the magic arrow,
> Staining all the waves with crimson
> With the crimson of its life-blood
> Filling all the air with splendour
> With the splendour of its plumage?

The story goes on to tell how the hunter speeds westward in pursuit of the Red Swan. "At lodges where he rests, they tell him she has often passed there, but those who followed her have never returned. She is the daughter of an old magician who has lost his scalp which Ojibwa succeeds in recovering for him and puts back on his head, and the old man rises from the earth no longer aged and decrepit but splendid in youthful glory. Ojibwa departs and the magician calls forth the beautiful maiden, now not his daughter, but his sister, and gives her to his victorious friend. It was in after days when Ojibwa had gone home with his bride that he travelled forth and coming to an opening in the earth descended and came to the abode of departed spirits, there he could behold the bright western region of the good, and the dark cloud of wickedness. But the spirits told him that his brethren at home were quarrelling for the possession of his wife, and at last after long wandering this Red Indian Odysseus returned to his mourning, constant Penelope, laid the magic arrows to his bow, and stretched the wicked suitors dead at his feet.

"Thus savage legends from Polynesia and America may well support the theory that Odysseus visiting the Elysian fields and Orpheus descending to the land of Hades to bring back the wide-shining Eurydikê are but the Sun himself descending to and ascending from the world below."

The Algonquin deity Manabozho was a personification of the sun, for, in an Ottawa myth, he is referred to as the elder brother of the Spirit of the West, God of the country of the dead in the region of the setting sun, and his solar character is further revealed in the legend of his vain pursuit of the West, his brother, to the brink of the world.

According to a Peruvian myth, Viracocha, the Supreme God of the Peruvians, rose from the bosom of Lake Titicaca, and journeying westward overcame all the foes that opposed him, and disappeared at length into the western sea, thus portraying his true solar character.

Faber[1] tells us that the ancient Mexicans believed that the world was made by the gods, but professed ignorance as to the precise mode in which it was formed. "They imagined that since the creation four suns had successively appeared and disappeared, and they maintained that that which we now behold is the fifth. The first sun perished by a deluge of water, and with it all living creatures. The second fell from heaven at a period when there were many giants in the country and by the fall everything that had life was again destroyed. The third was consumed by fire and the fourth was dissipated by a tempest of wind. At that time mankind did not perish as before but were changed into apes, yet when the fourth sun was blotted out there was a darkness which continued twenty-five years. At the end of the fifteenth

[1] *The Origin of Pagan Idolatry*, George Stanley Faber.

year their chief god formed a man and a woman who brought forth children, and at the end of the other ten years appeared the fifth sun then newly born. Three days after this last sun became visible all the former gods died, then in process of time were produced those whom they have since worshipped."

The Egyptians had a legend which in some respects is so similar to that of the Mexican myth related above that it would almost appear as if the two originated from the same source. They told Herodotus that, according to their records, the sun had four times deviated from his regular course, having twice risen in the west, and twice set in the east. This change, however, had produced no alteration in the climate of Egypt, neither had a greater prevalence of disease been the consequence.

Among the Maoris of New Zealand we find a myth that depicts dramatically the setting sun as it goes to its death through the western portals of the night. Because of its interest as a pronounced type of the solar myth it is given in detail:

"Maui, the New Zealand cosmic hero, at the end of his glorious career came back to his father's country and was told that here perhaps he might be overcome, for here dwelt his mighty ancestress 'Great-Woman-Night,' whom you may see flashing, and as it were opening and shutting there where the horizon meets the sky. What you see yonder shining so brightly red are her eyes, and her teeth are as sharp and hard as pieces of volcanic glass. Her body is like that of a man, and as for the pupils of her eyes they are jasper. Her hair is like the tangles of long seaweed, and her mouth is like that of a barracouta.

"Maui boasted of his former exploits, and said: 'Let us fearlessly seek whether men are to die or live forever.' But his father called to mind an evil omen that when he was baptising Maui he had left out part of the fitting prayers and therefore he knew that his son must perish, yet he said: 'O my last born, and the strength of my old age, . . . be bold, go and visit your great ancestress who flashes so fiercely there where the edge of the horizon meets the sky.' Then the birds came to Maui to be his companions in the enterprise, and it was evening when they went with him, and they came to the dwelling of his mighty ancestress, and found her fast asleep. Maui charged the birds not to laugh when they saw him creep into the old chieftainess, but when he had got altogether inside her, and was coming out of her mouth, then they might laugh long and loud. So Maui stripped off his clothes and crept in. The birds kept silence, but when he was in up to his waist the little tiwakawaka could hold its laughter no longer, and burst out loud with its merry note, then Maui's ancestress awoke, closed on him, and caught him tight and he was killed. Thus died Maui, and thus death came into the world.

"The New Zealanders hold that the sun descends at night into his cavern, bathes in the water of Life, and returns at dawn from the underworld; hence we may interpret their thought that if Man could likewise descend into Hades and return, his race would be immortal.

"It is seldom that solar characteristics are more distinctly marked in the several details of a myth than they are here. Great-Woman-Night who dwells on the horizon is the New Zealand Hades. The birds are to keep silence as the sun enters night, but may sing when he comes forth from her mouth, the mouth of Hades. The tiwakawaka describes the cry of the bird which is only heard at sunset."[1]

One of the most wide-spread and best-known sun myths relates to the devouring of the day by the night monster at set of sun, and the disgorging of the victim by the devourer in the morning. A Zulu legend describes the maw of this sun-devouring monster as a land teeming with human life,

[1] *Primitive Culture*, Edward B. Tylor.

and its environment, and when the monster is cut open, all the creatures issue forth from the state of darkness, the cock leading, exclaiming: "I see the world."

The well-known fairy tale of "Little Red Riding-Hood" is a sun myth of this type, and in Germany there is added to the tale the fact that, after the wolf had devoured his victim, a hunter slew the wolf, ripped him open, when out stepped the little maiden in her red cloak, safe and sound.

There is a legend current in Germany that relates to a frog that wooed the daughters of a queen. The youngest daughter consented to become his bride, and this gracious act on her part freed the frog from a magic spell, and he was transformed into a handsome youth.

"This tale," says Professor Max Müller,[1] "is solar in its character, and but another version of the Sanscrit story of Bhekî the frog who became the wife of a king only to vanish at the sight of a glass of water, a legend that grew out of a phrase which was possibly, 'the sun dies at the sight of water.'"

Another ancient myth that has come down to us relates to the mystic meeting of the sunlight and moonlight. The light of the sun was a king's daughter who, on a certain day asked to be allowed

[1] *Chips from a German Workshop,* Professor Max Müller.

to walk unattended in the streets of a great city.
The king consented, and ordered all the citizens to
remain indoors behind closed shutters on that day
and refrain from looking out. A minister, who
was really the moonlight, could not restrain his
curiosity. He stepped out on his balcony and
was seen by the king's daughter, who beckoned to
him and he joined her at the foot of a tree. Thus
did the sunlight and moonlight mingle their beams
of light. The king was told of their meeting and
set out for the trysting-place, but before his arrival
the minister's wife, realising her husband's peril,
sought him and so disguised him that he resembled
a monster. When the king finally found his
daughter, and saw no one near her but a monster,
he was convinced that he had been misinformed,
and that his daughter had met no one.

The association of the sun with a floating island
is revealed in many legends, and in solar symbolism
we find the sun depicted as seated on a floating
lotus leaf.

Herodotus tells us that near Buto there was a
deep and broad lake in which was a reputed floating island. In this island was a large temple
dedicated to the sun. The island was once firm,
but it is said when Typhon, who was the sea, was
once roaming round the world in pursuit of the
solar deity Horus, Latona, who was one of the

primitive eight gods who dwelt in the city of Buto, received him in trust from Isis and concealed him in the island of Chemmis, which then first began to float.   Afterwards he became sufficiently powerful to leave his place of refuge and to expel Typhon who had usurped his dominions, and his own reign then commenced.

The myth of Phœbus Apollo is substantially identical with this, and the island of Delos, the birthplace of the Sun-God, corresponds to the floating island Chemmis.

There is another parallel legend among the Peruvians.   When all mankind were swept away by the waters of the Deluge, a personage named Viracocha emerged from Lake Titicaca and became the founder of the sacred city of Cuzco.   Viracocha was the Sun-God of the Peruvians, and the common ancestor of the race of Incas.

In Lake Titicaca, which is considered sacred by the Peruvians, there is a small island where they claim the Sun-God hid himself and saved his life when the world was destroyed by the waters of the Deluge.   On this island there was a temple dedicated to the sun, as there was on the island of Chemmis, the Egyptian island, and the Greek island Delos.   These islands were considered holy places.

Faber[1] tells us: "The sun is further repre-

[1] *The Origin of Pagan Idolatry*, G. S. Faber.

sented as peculiarly delighting to haunt the sacred mountain which first raised its head above the retiring waters, and which received the ark. This mountain top, therefore, had the appearance of a floating island which doubtless gave rise to the many myths that represent the sun as navigating the deep.

"The favourite residence of the Greek solar deity was Parnassus. In the Zend Avesta the sun is described as ruling over the world from the top of Mount Albordi which is said to have been the first land that appeared above the waves of the retreating flood.

"The old Orphic poet, the priests of Egypt, and the Brahmas of Hindostan agree in maintaining that the sun was born out of an egg which had floated on the ocean, and which had been tossed about at the mercy of the elements."

The following extremely interesting solar myth of Irish extraction is related in *Myths and Myth Makers* by John Fiske:

"Long before the Danes ever came to Ireland, there died at Muskerry a Sculloge, or country farmer, who by dint of hard work and close economy had amassed enormous wealth. His only son did not resemble him. When the young Sculloge looked about the house, the day after his father's death, and saw the big chests full of gold

and silver, and the cupboards shining with piles
of sovereigns, and the old stockings stuffed with
large and small coin, he said to himself, 'Bedad,
how shall I ever be able to spend the likes o' that?'
And so he drank, and gambled, and wasted his
time in hunting and horse-racing, until after a
while he found the chests empty and the cupboards poverty-stricken, and the stockings lean
and penniless. Then he mortgaged his farm-
house and gambled away all the money he got for
it, and then he bethought him that a few hundred
pounds might be raised on his mill. But when he
went to look at it, he found the dam broken and
scarcely a thimbleful of water in the mill-race,
and the wheel rotten, and the thatch of the house
all gone, and the upper millstone lying flat on the
lower one, and a coat of dust and mould over
everything. So he made up his mind to borrow
a horse and take one more hunt to-morrow and
then reform his habits.

"As he was returning late in the evening from
his farewell hunt, passing through a lonely glen
he came upon an old man playing backgammon,
betting on his left hand against his right, and crying
and cursing because the right would win. 'Come
and bet with me,' said he to Sculloge. 'Faith, I
have but a sixpence in the world,' was the reply;
'but if you like, I'll wager that on the right.'

'Done,' said the old man, who was a Druid; 'if you win I'll give you a hundred guineas.' So the game was played, and the old man, whose right hand was always the winner, paid over the guineas and told Sculloge to go to the Devil with them.

"Instead of following this bit of advice, however, the young farmer went home and began to pay his debts, and next week he went to the glen and won another game, and made the Druid rebuild his mill. So Sculloge became prosperous again, and by and by he tried his luck a third time, and won a game played for a beautiful wife. The Druid sent her to his house the next morning before he was out of bed, and his servants came knocking at the door and crying, 'Wake up, wake up, Master Sculloge, there's a young lady here to see you.' 'Bedad, it's the vanithee¹ herself,' said Sculloge; and getting up in a hurry, he spent three-quarters of an hour in dressing himself. At last he went downstairs, and there on the sofa was the prettiest lady ever seen in Ireland. Naturally, Sculloge's heart beat fast and his voice trembled, as he begged the lady's pardon for this Druidic style of wooing, and besought her not to feel obliged to stay with him unless she really liked him. But the young lady, who was a king's daughter from a far country, was wondrously

¹ Lady of the house.

charmed with the handsome farmer, and so well did they get along that the priest was sent for without further delay, and they were married before sundown. Sabina was the vanithee's name; and she warned her husband to have no more dealings with Lassa Buaicht, the old man of the glen. So for a while all went happily, and the Druidic bride was as good as she was beautiful. But by and by Sculloge began to think he was not earning money fast enough. He could not bear to see his wife's hands soiled with work, and thought it would be a fine thing if he could only afford to keep a few more servants, and drive about with Sabina in an elegant carriage, and see her clothed in silk and adorned with jewels.

" 'I will play one more game and set the stakes high,' said Sculloge to himself one evening, as he sat pondering over these things; and so, without consulting Sabina, he stole away to the glen, and played a game for ten thousand guineas. But the evil Devil was now ready to pounce on his prey, and he did not play as of old. Sculloge broke into a cold sweat with agony and terror as he saw the left hand win. Then the face of Lassa Buaicht grew dark and stern, and he laid on Sculloge the curse which is laid upon the solar hero in misfortune, that he should never sleep twice under the same roof, or ascend the couch of the dawn-nymph, his wife, until he should have procured and brought to him the sword of light. When Sculloge reached home, more dead than alive, he saw that his wife knew all. Bitterly they wept together, but she told him that with courage all might be set right. She gave him a Druidic horse, which bore him swiftly over land and sea, like the enchanted steed of the Arabian Nights, until he reached the castle of his wife's father, who, as Sculloge now learned, was a good Druid, the brother of the evil Lassa Buaicht. This good Druid told him that the sword of light was kept by a third brother, the powerful magician Fiach O'Duda, who dwelt in an enchanted castle, which many brave heroes had tried to enter, but the dark sorcerer had slain them all. Three high walls surrounded the castle, and many had scaled the first of these, but none had ever returned alive. But Sculloge was not to be daunted, and taking from his father-in-law a black steed, he set out for the fortress of Fiach O'Duda. Over the first high wall nimbly leaped the magic horse, and Sculloge called aloud to the Druid to come out and surrender his sword. Then came out a tall, dark man, with coal-black eyes and hair and melancholy visage, and made a furious sweep at Sculloge with the flaming blade. But the Druidic beast sprang back over the wall in the twinkling of an eye and rescued his rider, leaving, however, his tail behind in the court-yard. Then Sculloge returned in triumph to his father-in-law's palace, and the night was spent in feasting and revelry.

"Next day Sculloge rode out on a white horse, and when he got to Fiach's castle, he saw the first wall lying in rubbish. He leaped the second, and the same scene occurred as the day before, save that the horse escaped unharmed. The third day Sculloge went out on foot, with a harp like that of Orpheus in his hand, and as he swept its strings the grass bent to listen and the trees bowed their heads. The castle walls all lay in ruins, and Sculloge made his way unhindered to the upper room, where Fiach lay in Druidic slumber, lulled by the harp. He seized the sword of light, which was hung by the chimney sheathed in a dark scabbard, and making the best of his way back to the good king's palace, mounted his wife's steed, and scoured over land and sea until he found himself in the gloomy glen where Lassa Buaicht was still crying and cursing and betting on his left hand against his right.

" 'Here, treacherous friend, take your sword of light'; shouted Sculloge in tones of thunder; and as he drew it from its sheath the whole valley was lighted up as with the morning sun, and next moment the head of the wretched Druid was lying at his feet, and his sweet wife, who had come to meet him, was laughing and crying in his arms."

Some authorities claim that the legend of William Tell is a sun myth. He is admittedly a skilful navigator, a practised archer, and, as the myth relates, after he had successfully emerged from the storm and tempest he leaps at dawn, rejoicing in his freedom on the land, and slays the tyrant who had enslaved him. These facts are all well in accord with those predominating in the typical solar myth.

It is quite impossible in a volume treating of sun lore in all its phases to discuss the solar myth exhaustively. An attempt has been made to indicate that primitive man, wherever he was situated, strove to interpret natural phenomena in the familiar language of his daily existence, and to attribute to the manifestations of physical laws a human agency. The result of this close observation of nature led to the deification of its powers and paved the way for a wealth of imagery, a treasure of myth and legend, which lends colour and brightness to the more sombre pages of the early history of man.

Chapter V
Solar Folk-Lore

Chapter V

Solar Folk-Lore

THE distinction between mythology and folk-lore is an extremely fine one, and though there is such a distinction, still the two subjects are so essentially analogous it will not be strange if portions of the material in this chapter would, according to some authorities, seem misplaced, and more properly included in the chapter on Solar Mythology, and *vice versa*. In view of the difficulties of an absolutely correct classification, the author makes no claim that his is the correct one.

In the early stages of the history of man, every act of nature and the movements of the heavenly bodies was attributed to the machinations of some one, a mysterious personage, an all-powerful being, an unseen god. The sun, as the chief luminary, commanded man's attention from the earliest days, and it was but natural for primitive man to speculate on the phenomena of his daily appearance and disappearance in terms that seem to us now childish and puerile.

To men who looked to the west across a vast expanse of sea, the sun at nightfall seemed to sink directly into the waves, and, as they were confident that the sun was an extremely large and hot body, they were convinced it would give out a hissing noise when the waters closed over it.

From the expression of the thought to the actual fact was but a step, and so we find Posidonius recording that the inhabitants of Cape St. Vincent, the westernmost point of Europe, claimed that the sun disappears each night into the sea with a great hissing noise.

We find the same idea current in the islands of Polynesia, in Iberia, and Germany, where the people claim to have heard the mighty hissing of the sea-quenched sun.

The Egyptians regarded the sun as a child when it was rising, and as an old man when it was setting in the evening. These ideas were also transferred to the annual motion of the sun. Macrobius states that the Egyptians compared the yearly course of the sun with the phases of human life; thus, a little child signified the winter solstice, a young man the spring equinox, a bearded man the summer solstice, and an old man the autumnal equinox. They also thought that Hercules had his seat in the sun, and that he travelled with it round the moon.

The Hindus often referred to the sun as "the eye of Mithra, Varuna, and Agni," and at sunrise or sunset, when the sun appeared to be squatting on the water, they likened it to a frog. This simile gave rise to a Sanscrit story, which is found also in German and Gaelic.

"Bhekî (the frog) was a beautiful maiden. One day when she was sitting near a well, a king rode by, and fascinated by her beauty, asked her hand in marriage. She consented on the condition that he would never show her a drop of water. He accepted, and they were married. One day being tired and thirsty she asked the king for a glass of water, and forgetting his promise, he granted her request, and his bride immediately vanished. That is to say, the sun disappeared when it touched the water."

The sun was also regarded as a well, and in the Semitic, Persian, and Chinese languages the words "well" and "eye" are synonymous. Considered as a well, the rays of the sun were likened to the moisture that flows from the well.

In different parts of Africa we find the sun variously regarded. In Central Africa, where it is extremely hot, the rising of the sun is always dreaded, and the orb of day is a common enemy. It was the custom, among certain tribes, to curse the sun at his rising for afflicting the people with burning heat.

In Southern Africa, on the contrary, the natives believed that they were descended from the Sun; and if, by chance, the rising of the sun was obscured by clouds, they thought the Sun purposely hid his face from them because their misdeeds offended him, and straightway they performed acts of propitiation. Work at once ceased, and the food of the previous day was given to the old women. The men of the tribe then went in a body to the river to purify themselves by washing in the stream. Each man threw into the river a stone from his hearth, and replaced it with a new one from the bed of the river. On returning to the village the chief kindled a fire in his hut, and the members of the tribe all gathered embers from it to light their individual hearth fires. The ceremony concluded with a dance in which the whole tribe joined. The idea seems to have been, that the lighting of the flame on earth would serve to rekindle the dead solar fire. When the sun set, these people said "The Sun dies."

The early inhabitants of Polynesia called the sun "Ra," which was also the Egyptian sun name. They believed that it was endowed with life, and the offspring of the gods. To account for its rise in the east each morning, after its disappearance in the west each night, they said that during the night it passed through a passage under the seas, so as to rise in its appointed place in the eastern sky each day.

In some of the islands the sun was thought to be a substance resembling fire, and they regarded its disappearance each night as a falling of the orb into the sea, and, as we have seen, the inhabitants of the westernmost islands were confident that they had heard the hissing occasioned by the sun's plunge into the ocean.

The early tribes seemed to think they could control the light of the sun and stay or hasten its setting. "The Melanesians make sunshine by means of a mock sun," says Frazer.[1] "A circular stone is wound about with red braid and stuck with owl's feathers to represent the rays of the sun, or the stone is laid on the ground with white rods radiating from it to imitate sunbeams." A white or red pig is sacrificed in the sunshine-making ceremony, and a black one when rain is desired.

In New Caledonia they burnt a skeleton to make sunshine, and drenched it with water if they wished for rain. They also had a more elaborate ceremony for producing sunshine, which Frazer[2] thus describes: "When a wizard desires to make sunshine he takes some plants and corals to the burial ground, and makes then into a bundle, adding two locks of hair cut from the head of a living child

[1] *The Golden Bough*, J. G. Frazer.   [2] *Ibid*.

(his own child if possible), also two teeth, or an entire jawbone from the skeleton of an ancestor. He then climbs a high mountain whose top catches the first rays of the morning sun.   Here he deposits three sorts of plants on a flat stone, places a branch of dry coral beside them, and hangs the bundle of charms over the stone.   Next morning he returns to this rude altar, and at the moment when the sun rises from the sea, he kindles a fire on the altar.   As the smoke rises he rubs the stone with the dry coral, invokes his ancestors and says: 'Sun: I do this that you may be burning hot, and eat up all the clouds in the sky.' The same ceremony is repeated at sunset.''

The sun, according to many traditions of primitive man, spent a part of its time in the underworld, or in a submarine passage beneath the seas, and if it did not go of its own volition, it was carried there by some enemy.   Thus in Servia a tale is told, that when the devils fell, their king carried off the sun from heaven affixed to a lance. This was a great calamity, and the Archangel St. Michael was selected to try to recover it.   He therefore set out for the underworld and succeeded in making friends with the archfiend.   As they stood together by a lake, St. Michael proposed to the devil that they engage in a diving contest. The latter consented, and thrusting the lance which held the sun into the ground, he dived in. This was St. Michael's opportunity, and making the sign of the cross, he grabbed the lance and made off, hotly pursued by the Evil One. Being fleet of foot he outdistanced him, but his pursuer was so close to him at one time that he managed to scratch his foot. In honour of St. Michael and his valiant deed, men, from that time on, were destined to have indented soles.

The old Germans called the sun "Wuotan's eye," and there is a German legend that reveals the sun as the punisher of evil thinkers: It appears that a prisoner was once on his way to execution, an object of pity to all whom he passed, but one woman, who was engaged in hanging up her linen to dry in the sun, remarked that he well deserved his fate. Immediately her linen fell to the ground, nor was she able to hang it up in this drying-place thereafter. It is further related that, at her death, she was taken up to the sun to remain there as long as the world endures, as a punishment for her lack of pity.

The peasants in various parts of Germany call the Milky Way the "Mealway" or the "Millway," and say that it turns with the sun, for it first becomes visible at the point where the sun has set. It leads, therefore, to the heavenly mill, and its colour is that of the meal with which it is strewed. This brings us to the Norse story of "The Wonderful Mill,"[1] an exceedingly interesting bit of folk-lore of solar significance. "The peasants of Norway to this day tell of the wondrous mill that ground whatever was demanded of it. The tradition is of great antiquity. The earliest version known is as follows: Of all beliefs, that in which man has at all times of his history been most prone to set faith, is that of a golden age of peace and plenty which has passed away, but which might be expected to return. Such a period of peace and plenty, such a golden time, the Norsemen could tell of in his mythic Frodi's reign, when gold, or Frodi's meal, as it was called, was so plentiful that golden armlets lay untouched from year's end to year's end on the King's highway, and the fields bore crops unsown. In Frodi's house were two maidens of that old giant race, Frenja and Menja. These daughters of the giant he had bought as slaves, and he made them grind his quern or hand-mill Grotti, out of which he used to grind peace and gold. Even in that golden age one sees there were slaves, and Frodi, however bountiful to his thanes and people, was a hard taskmaster to his giant handmaidens. He kept them to the mill, nor gave them longer rest than the cuckoo's note lasted, or they could sing a song. But that quern

[1] From Dasent's *Popular Tales from the Norse.*

was such that it ground everything that the grinder chose, though until then it had ground nothing but gold and peace. So the maidens ground and ground, and one sang their piteous tale in a strain worthy of Æschylus, as the other rested. They prayed for rest and pity, but Frodi was deaf. Then they turned in giant mood, and ground no longer peace and plenty, but fire and war. Then the quern went fast and furious, and that very night came Mysing the sea-rover and slew Frodi and all his men, and carried off the quern, and so Frodi's peace ended. The maidens, the sea-rover took with him, and when he got on the high seas he bade them grind salt, so they ground, and at midnight they asked if he had not salt enough, but he bade them grind on. So they ground till the ship was full and sank. Mysing, maids, mill, and all, and that's why the sea is salt."

This wonder-working mill once stood in heaven, it is said, for Frodi its owner was no other than the Sun-God Freyr. The flat circular stone of Frodi's quern is the disk of the sun, and its handle is the pramantha with which Indra or the Aswins used to kindle the extinguished luminary.

To explain the circular motion of the sun, the Incas of Peru believed that it was hung in space by a cord, and that each evening it entered the sea, and being a good swimmer it pierced through the waves, and reappeared next morning in the east.

The Incas claimed that the Sun was their own elder brother, and ruled over the cohorts of heaven by divine right. Their legends relate that the Sun took pity on the children of men, who, in primitive times, lived in a state of savagery, and he therefore sent his son and daughter to enlighten them, and teach them to live properly. They are said to have risen from the depths of Lake Titicaca, that marvellous sheet of water twelve thousand feet above the sea. They taught the Peruvians the essentials of culture and education.

According to another tradition, the Peruvians traced their origin from the first Inca, the Sun and his wife, who came from the island of the sun in Lake Titicaca, and founded the city of Cuzco, the sacred city of the sun. This island in the lake is therefore sacred to the Peruvians, and many ruins of the Incas are to be found there.

The Peruvians paid particular attention to the daily meridian passage of the sun, and observed that when it was in the zenith, it cast no shadow.

The early natives of Brazil believed that the sun was a ball of light feathers, which some mysterious being exhibits during the day, and covers at night with a pot.

The folk-lore of the North American Indian

Isle of Sun, Lake Titicaca, Peru

Courtesy of Mr. Leon Campbell

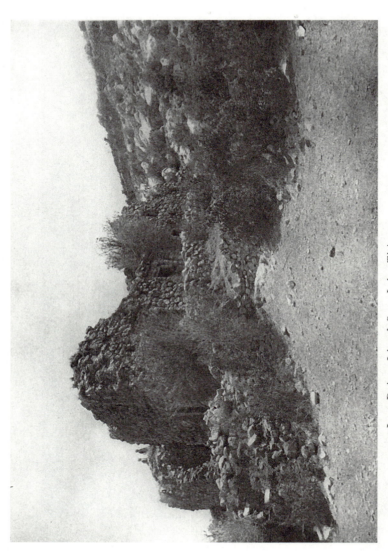

Inca Ruins, Isle of Sun, Lake Titicaca

Courtesy of Mr. Leon Campbell

tribes is rich in legends respecting the sun. The Indians believed that the sun was an animated being endowed with human attributes. The following tales are related by the Thompson River Indians:

There was once a most mischievous and incorrigible youth who one morning strolled away from his home. On his return, he found that his parents had deserted him, but his old grandmother, who was unable to travel, was left behind. She taught the boy how to make a bow and arrows, and with these he was able to provide a daily supply of food. She also made blankets for him out of the skins of many coloured birds. These were of such beauty that they attracted the attention of the Sun. It had always been the custom of the Sun to travel about naked during the day, and clothe himself only in the dark hours.[1] But when the Sun saw these beautiful blankets, he purchased them from the boy, and wrapped them about his body, and soon disappeared, so that at set of sun you may see the gorgeous colouring of these robes in the western sky, especially the blue tint of the blue-jay blanket.

Another tale relates that originally the Sun lived much nearer the earth than now,[2] and preyed

[1] This may have been the Indian way of accounting for the invisibility of the sun at night.

[2] It is strange that the nebular hypothesis conforms with this idea, that the sun and earth were close together at one time.

upon mankind. It was his custom to kill people every day on his travels, and carry them off to his home at night-fall to eat. His son lived quietly at home clad in fine garments of many colours. There was once an Indian who in gambling was most unlucky. One day, while much depressed, he set out on a journey in search of adventure, and finally came to the Sun's abode in the absence of the owner. The son received him kindly, but fearing that his guest would be discovered by his cannibal father, he hid him under a heap of robes. The Sun arrived in the evening carrying a man on his back, and as he came near the house, he said: "Mum, Mum, Mum.[1] There must be a man here," but his son persuaded him that he was mistaken. The next day the Indian was glad to leave this dangerous locality, and returned to his home laden with gifts from his benefactor. Out of gratitude he returned later to the Sun's house and made his friend the present of a wife and one for his father. This pleased the Sun so much that he gave up the killing and eating of human beings. In the foregoing legend we find expressed the idea, current in the traditions of many primitive people, that celestial beings feed on human bodies.

[1] We are almost tempted to add, "I smell the blood of an Englishman," for here we have a tale identical in many particulars with the popular fairy tale of "Jack the Giant Killer," which some authorities claim is of solar origin.

The following tale is told of the Sun and his daughter:[1]

Originally the Sun was an eminent chief, possessed of great power and wealth. He was also blessed with a beautiful daughter, and the fame of her beauty spread afar. A powerful magician, entranced with the maiden, sought her hand, and though at first repulsed, finally won the Sun's favour and married his daughter. The Sun implored his daughter to visit him frequently. This, however, she neglected to do, and, finally, when she did go to her father with her two children, he transformed her into the present Sun. This is why the Sun travels each day from east to west in search of her father. Her children are occasionally seen as sun-dogs closely following their mother.

The Indians of Northern California relate the following story:

Once the sun fell by accident down from the sky just about sunrise, but the quick little mole was watching, and caught it before it touched the earth, and succeeded in holding it up until others arrived, when, by exerting all their strength, they

---

[1] This myth is typical of many that may well be styled Evaporation and Rainfall myths that are thus interpreted. The water is enamoured of the cloud, the beautiful daughter of the Sun. The Sun does not favour the suitor, and strives to kill him by subjecting him to a number of tests. The Water achieves success in all of these, and then receives the Sun's permission to marry his daughter.

succeeded in replacing it where it belonged in the sky, but ever thereafter the mole's hands were bent far back to show how he had worked to hold up the sun.

As evidence of the Indian belief in the Sun's solicitude in their affairs, and his protecting and saving influences, the Cheyenne tale of "The Eagle Hunter" is told:

There was an Indian who once set out to catch an eagle. Digging a hole in the ground he crept in, covered it over with brush, and cleverly baited it with a skinned buffalo calf. Presently an eagle espied the prey, flew down, and began to eat of it, when the Indian laid hold of its feet, and held it captive; but he had underestimated the power of the bird, which had strength enough to carry the man up to a mountain crag, where it was impossible for him to descend. The Indian realising his desperate plight, prayed to the Sun for deliverance, and the Sun, taking pity on him, sent a great whirlwind which swept the hunter from his lofty perch, and safely deposited him on the ground.

In a Maidu legend it is related that the Sun dwells in an impregnable house of ice into which she retreats after killing people on the earth. Once she abducted the Frog's children, and was closely pursued by their angry mother, who finally overtook the Sun and swallowed her, but the Sun

Ruins of the Incas' Bath, Isle of Sun, Lake Titicaca

Courtesy of Mr. Leon Campbell

The Temple of Jupiter, Baalbek, Syria

burst her open and transformed her into a Frog again.

There are many Indian tales wherein the sun figures as a target. The Shoshone Indians believed that, in the beginning, the sun did not shine till the Rabbit shot at him with his magical arrow (the fire drill).

In the following Mewan Indian legend,[1] the sunlight is extinguished by the arrow shot: "There was once a poor worthless Indian boy who got his living by begging. At length, finding people loath to assist him, he threatened to shoot out the sun, and as this had no effect, he made good his threat, and shot the sun, thus letting its light out, and the whole world became dark. It was dark for years, and every one was starving for want of light, when the Coyote-Man discovered a dim light a long distance off, and sent the Humming-bird to investigate. The bird, finding its way to the sun, pecked off a piece, and returned with it under its chin, and making repeated trips finally succeeded in restoring the full light of the sun, and to this day you can see the marks of its burden beneath the chin of the Humming-bird." This association of the Humming-bird with the sun is found in the traditions of the Aztecs. In their temples was enthroned a deity known as "the Humming-bird

[1] *The Dawn of the World*, C. Hart Merriam.

to the left," and this bird was considered by them to be a divine being, the emissary of the sun. In the Aztec language it is often called "Sunbeam," or "Sun's hair."

Among the Indians there seems to have been an almost universal tradition that originally men lived in a world of darkness, or semi-darkness, before the sun was placed in the heavens. A Mewan legend relates that, in the early days, the land was shrouded in fog, and was cold and dark. It was such a poor place to live in that Coyote-Man was not satisfied with the conditions, and set out on a journey to seek some way to better it. He finally came to a pleasant land of sunshine, and, charmed with it, returned to tell his people of the delightful land he had visited. They suggested that he offer to buy the sun, so he returned to the land of light and made this proposition, but it was rejected, so Coyote-Man resolved to steal the sun as his people were in sore need of it.

This was a difficult matter as the sun was carefully watched by the Turtle, who slept with one eye always open. Coyote-Man, resorting to magic, took the form of a big oak log, and the Turtle, when out seeking for wood, took him and threw him on the fire. But the fire did not even singe him, and seeing the Turtle asleep, he resumed his form, seized the sun and ran off with it to his own land.

The people, however, did not understand it, and bade Coyote-Man make it go, and, as he was sorry for the people he had deprived of the sun, he arranged a plan so that the sun could light up both lands. He carried the sun west to the place where the sky joins the earth, and found the place for the sun to crawl through, and where it could go down under the earth, and come up in the eastern sky in the morning through the hole in the east. The sun did his bidding, and thus both lands thereafter rejoiced in the blessedness of sunshine.

The Natchez of Mississippi, the Apalachees of Florida, the Mexicans and Peruvians, all believed that the sun is the bright dwelling-place of their departed chiefs and warriors.

A primitive Mexican prayer offered in time of war embodies this idea: "Be pleased, O our Lord, that the nobles who shall die in the war be peacefully and joyously received by the sun and the earth, who are the loving father and mother of all."

It is said that General Harrison once called the Shawnee chief, Tecumseh, for a conference: "Come here, Tecumseh, and sit by your father," he said. "You my father?" replied the chief with a stern air, "No, yonder sun [pointing toward it] is my father, and the earth is my mother, so I will rest on her bosom," and he sat upon the ground.

The Kootenay Indians speak of the sun as a

blind man who is cured by his father-in-law, Coyote-Man. Here we have another reference to the Coyote's service to mankind in bringing sunshine to his people.

Among the New Zealanders the sun is regarded as a great beast whom the hunters thrashed with clubs. His blood is supposed to be used in some of their incantations, and according to an Egyptian tradition, the sun's blood was kneaded into clay at the making of man.

We have seen how the sun was metaphorically regarded in India and other lands not merely as a human creature, but as the eye of a supreme and all-seeing deity. In like manner the inhabitants of Java, Sumatra, and Madagascar called the sun "the eye of day." This metaphor has been used extensively even in modern poetry.

When the astronomers Galileo, Scheiner, and Fabricius discovered the spots on the sun, the Aristotelians indignantly insisted that they were mistaken, and that the phenomenon was due to defects in the optical properties of their telescopes or eyes. They argued that it was quite incompatible with the dignity of the Eye of the Universe that it should be afflicted with such a common ailment as ophthalmia.

Tylor[1] tells us that the Rev. Tobias Snowden,

[1] *Primitive Culture*, Edward B. Tylor.

in a book published in the last century, proved the sun to be Hell, and the dark spots, gatherings of damned souls.

In Greece there was a general protest when the astronomers denied not only the divinity, but the very personality of the sun, and declared it to be nothing but a huge fiery globe. These statements were regarded as blasphemous, and, in fact, Anax-agoras was punished with death for having taught that the sun was not animated, and that it was nothing but a mass of iron, about the size of the Peloponnesus.

Such a state of affairs strikes us in this enlightened age as decidedly extraordinary, and yet in the history of the early settlers of this country we have in the trials for witchcraft an equally absurd and foolish state of affairs.

Every age, therefore, to be judged fairly on its merits, must be viewed in the light of its state of progress, and, grotesque as many of the foregoing legends related of the sun may seem, it behooves us to withhold our mirth, and endeavour to realise how much these traditions were a serious part of the lives of the people of unenlightened ages.

Chapter VI
Sun Worship

Chapter VI

Sun Worship

THE pre-eminence of the Sun, as the fountain-head of life and man's well-being, must have rendered it at a date almost contemporaneous with the birth of the race, the chief object of man's worship.

"It was," says Kames,[1] "of all the different objects of idolatry the most excusable, for upon the sun depend health, vigour, and cheerfulness, and during its retirement all is dark and disconsolate." Hence, as we shall see, the chief masculine deity of every nation which was the chief object of their idolatrous worship, is in every case to be identified with the sun.

The Abbé Banier wrote in like vein:[2] "Nothing was more capable of seducing men than the Heavenly Bodies, and the sun especially. His beauty, the bright splendour of his beams, the rapidity of his course, *exultavit ut Gigus ad currendam viam,*

---

[1] *History of Man*, Hon. Henry Home of Kames.
[2] *The Mythology and Fables of the Ancients*, Abbé Banier.

his regularity in enlightening the whole earth by turns, and in diffusing Light and Fertility all around, essential characters of the Divinity who is Himself the light and source of everything that exists, all these were but too capable of impressing the gross minds of men with a belief that there was no other God but the sun, and that this splendid luminary was the throne of the Divinity. God had fixed his habitation in the heavens, and they saw nothing that bore more marks of Divinity than the sun." In the words of Diodorus Siculus: "Men in earlier times struck with the beauty of the Universe, with the splendour and regularity which everywhere were in evidence, made no doubt that there was some Divinity who therein presided, and they adored the sun as expressing the likeness of the Deity."

The worship of the sun was inevitable, and its deification was the source of all idolatry in every part of the world. It was sunrise that inspired the first prayers uttered by man, calling him to acts of devotion, bidding him raise an altar and kindle sacrificial flames.

Before the Sun's all-glorious shrine the first men knelt and raised their voices in praise and supplication, fully confirmed in the belief that their prayers were heard and answered.

Nothing proves so much the antiquity of solar

idolatry as the care Moses took to prohibit it.
"Take care," said he to the Israelites, "lest when
you lift up your eyes to Heaven and see the sun,
the moon, and all the stars, you be seduced and
drawn away to pay worship and adoration to the
creatures which the Lord your God has made for
the service of all the nations under Heaven."
Then we have the mention of Josiah taking away
the horses that the king of Judah had given to
the sun, and burning the chariot of the sun with
fire. These references agree perfectly with the
recognition in Palmyra of the Lord Sun, Baal
Shemesh, and with the identification of the
Assyrian Bel, and the Tyrian Baal with the sun.

Again, we have good evidence of the antiquity
of Sun worship in the fact that the earliest authentic
date that has been handed down to us was inscribed on the foundation stone of the temple of
the Sun-God at Sippara in Babylon by Naram-Sin,
son of Sargon. There has also been recovered an
ancient tablet, an inscribed memorial of the reign
of one of the early kings of Babylon, on which is
sculptured a representation of the worship of the
Sun-God by the king and his attendants. In the
sculpture, the Sun-God appears seated on a throne
beneath an open canopy shrine. He has a long
beard and streaming hair, like most conceptions of
the Sun-God, and in his hand he holds a ring, the

emblem of time, and a short stick too small for a sceptre, which some archæologists think represents the fire-stick which was so closely associated with the Sun-God. On a small table-altar, which stands before him, is a large disk ornamented with four star-like limbs, and four sets of wave-like rays, while above the group is the inscription: "The Disk of the Sun-God, and the rays (of his) eyes."

The scene clearly indicates the fact that the priests of Sippara were worshippers of the solar disk, and solar rays, and their creed seems to bear a close resemblance to that in vogue in the 18th Egyptian dynasty. The inscriptions on this memorial tablet are a valuable record of the religious life and ceremonial of the Babylonian temples.

The Babylonians, whose deity Shamash, the Sun-God, was worshipped at Sippara and Larsa, believed that in the firmament there were two doors—one in the east, and the other in the west. These were used by the Sun-God in his daily journey across the sky. He entered through the eastern door, and made his exit through the western portal. One tradition records that he rode in a chariot on his daily course drawn by two spirited horses.

In the representations of the Sun-God on the ancient cylinder seals, however, he is generally depicted journeying on foot. Each evening when the Sun-God disappeared in the west, he feasted and rested from his exertions in the abode of the gods, the underworld.

The authorities do not agree as to the place where the worship of the Sun was introduced, but perhaps those who claim Chaldea as the birthplace of Sun worship have the best of the argument, as it is well known that the Chaldeans were the first who observed the motion of the heavenly bodies, and astrology flourished in this reign in the earliest times. The principal deities worshipped by the Chaldeans were arranged in triads of greater and less dignity, nearly all the members of these being personifications of the heavens or the heavenly bodies.

The first triad comprised Ana, the heavens, or the hidden Sun, Father of Gods, Lord of Darkness, Lord of Spirits. Next in order came Bil, also a Sun-God, the Ruler, the Lord, the source of kingly power. His name has the same significance as Baal, and he personifies the same aspect of nature, the Sun ruling in the heavens.

The gods of the Canaanite nations, Moloch, Baal, Chemosh, Baalzebub, and Thammuz, were all personifications of the sun or the sun's rays, considered under one aspect or another. These cruel gods, to whom human sacrifices were offered, represented the strong fierce summer sun.

Solar worship was the predominant feature of the religion of the Phœnicians, and the source of their mythology. Baal and Ashtoreth, their chief divinities, were unquestionably the Sun and Moon, and a great festival in honour of the Sun-God, called "the awakening of Herakles," was held annually at Tyre, in February and March, representing the returning power of the Sun in spring. The Phœnician Sun-God, Melkarth, belonged to the line of Bel or Baal, and was the tutelary divinity of the powerful city of Tyre. Melkarth personified the Sun of spring, gradually growing more and more powerful as it mounts to the skies; hence the Phœnicians regarded him as a god of the harvests, and of the table, the god who brings joy in his train. Quails were offered as sacrifices at his altars, and as it was supposed that he presided over dreams, the sick and infirm were sent to sleep in his temples that they might receive in their dreams some premonition of their approaching recovery. The white poplar was particularly dedicated to his service. His votaries celebrated his worship with fanatical rites, invoking him with loud cries, and cutting themselves with knives. Strangely enough, in the North American Indian worship of the sun, a similar custom of self-mutilation is undergone in the sun-dance ceremonial.

The hardy Tyrian navigators soon spread this

solar worship from island to island even as far as Gades, where a flame burned continually in his temples. His name signifies, according to some, "the King of the City" or "the powerful King."

The Phœnicians also adored the Supreme Being under the name of Bel-Samen, and it is a remarkable fact that the Irish peasants have a custom, when wishing a person good luck, to say, "the blessing of Bel, and the blessing of Sam-hain be with you," that is, of the Sun and Moon.

The Israelites found the worship of Baal already prevailing in the interior of Palestine, and the adjacent countries on the east, when they came out of Egypt.

We know little of the ritualistic worship of Baal save that high places and groves were especially devoted to his honour, and regarded as sacred. He had a numerous priesthood, and a passage in Jeremiah reveals that human beings were sacrificed to his worship.

The ancient Persians were Sun worshippers, and Mithras, their Supreme Deity, represented the orb of day. Among these people, however, Fire worship soon became the predominating religion, which flourished under the guiding hand of Zoroaster.

The early inhabitants of Armenia likewise worshipped the Sun, and on festival occasions they were wont to sacrifice a horse to the object of their worship.

One of the most interesting evidences of ancient Sun worship has been brought to light in Syria, during the last few years, by German excavators who have been engaged in exposing the wonderful and imposing ruins at Baalbeck. Chief among these in importance is the Great Temple of the Sun, dedicated to Jupiter, and identified with Baal and the Sun. With him were associated both Venus and Mercury, under whose triple protection the ancient city of Heliopolis was placed. Unfortunately, the Great Temple has been almost entirely destroyed. All that remains are six columns of the peristyle, capped with Corinthian capitals, and joined by an elaborately decorated and massive entablature. An inscription on the great portico states that the temple was erected to the Great Gods of Heliopolis by Antonius.

In Egypt we find Sun worship exalted to the highest degree, and the Sun may well be regarded as the central object of the Egyptian religion. Diodorus says: "The first generation of men in Egypt, contemplating the beauty of the superior world, and admiring with astonishment the frame and order of the universe, imagined that there were two chief gods, eternal and primary, the Sun and Moon, the first of whom they called 'Osiris,' the other 'Isis.' They held that these gods governed the whole world cherishing and increasing all things."

The Egyptian priests taught that all their great deities were once men, but that after they died their souls migrated into some one or other of the heavenly bodies. As Osiris was declared to be the Sun, it is evident that, according to this system, the soul of the man was thought to have been translated into the solar orb, so that "when" says Faber,[1] "the pagans worshipped the sun as their principal divinity they did not worship him simply and absolutely as the mere chief of the heavenly luminaries, but they adored in conjunction with him and perpetually distinguished by his name the patriarch Noah, whose soul after death they feigned to have migrated into that orb, and to have become the intellectual regent of it. The person worshipped in the sun was not simply Noah but Noah viewed as a transmigratory reappearance of Adam. The setting and rising of the sun really meant the entrance into and the quitting of the Ark, or his death and resurrection."

The setting of the sun in the west at night, and its rising again in the east the following morning, presented a mystery to which the Egyptians attached great importance. To them the dis-

[1] *The Origin of Pagan Idolatry*, George Stanley Faber.

appearance of the sun signified the end of a contest,
the Sun-God vanquished by the demons of the
darkness, descended to the realm of death. "In
the *Book of the Dead*" says Tylor,[1] "it is written
that the departed souls descend with the Sun-God
through the western gate and travel with him
among the fields and rivers of the underworld."

The war that the Sun waged with his enemies
did not, however, end with his disappearance in
the west at eve of day. All through the hours of
darkness the battle went on in the underworld,
until, finally, the sun gained the upper hand,
and emerged victorious in the east, all glorious
and triumphant to gladden the hearts of men, and
proclaim the immortality of his soul; for, to the
Egyptians, the soul was wholly identified with
the Sun-God, and partook of all the vicissitudes
that befell him. It died with him at nightfall,
fought with him against the powers of darkness in
the underworld, and renewed its life with him in
the glories of the dawn.

The Egyptians, in the deification of the sun,
considered the luminary in its different aspects,
separating the light from the heat of the sun, and
the orb from the rays. Egyptian Sun worship
was therefore polytheistic, and several distinct
deities were worshipped as Sun-Gods. Thus,

[1] *Primitive Culture*, Edward B. Tylor.

there were Sun-Gods representing the physical orb, the intellectual Sun, the sun considered as the source of heat, and the source of light, the power of the sun, the sun in the firmament, and the Sun in his resting-place.

It is quite impossible in a work of this nature to adequately treat the subject of Egyptian Sun worship. Volumes have been written on the the subject and space forbids more than a brief account of the worship of the more important of the Egyptian solar deities.

The worship of the Sun-Gods Ra and Osiris was the most ancient religion mentioned on the oldest monuments of Egypt. "They are those," says Tiele,[1] "which in after times prevailed most generally and may be said to have formed the foundation of the national religion."

Undoubtedly the most important of the Egyptian Sun-Gods was Ra, and there appear to have been few gods in Egypt who were not at one time or another identified with him. As far back as Egyptian history reaches this Sun-God appears, as where in the pictures on the mummy cases, Ra, the Sun, is seen travelling in his boat through the upper and lower regions of the Universe, and his worship appears to have been universal throughout Egypt.

[1] *History of the Egyptian Religion*, Dr. C. P. Tiele.

Ra personified the physical sun, the glorious mid-day sun ruling the firmament, and symbolised to the ancient Egyptians the majesty and power of kings. He was worshipped as an omnipotent and all-powerful god under the names Ra and Amen-Ra.[1]

Wilkinson[2] tells us that the name of this deity was pronounced Rä, and with the definite article Pi prefixed it was the same as Phrah, or, as we erroneously call it, Pharaoh of the Scriptures. The Hebrew word Phrah is no other than the Memphitic name of the sun.

The hawk and globe emblems of the sun are placed over the banners or the figures of the Egyptian kings in the sculptures to denote this title. This adoption of the name of the sun as a regal title was probably owing to the idea that, as the Sun was the chief of the heavenly bodies, he was a fit emblem of the king who was the ruler over all the earth. In many of the kingly titles the phonetic nomen commenced with the name of Ra, as the Rameses, and others, and the expression "living forever like the sun, the splendid Phrê," are common on all the obelisks and dedicatory inscriptions.

[1] It is a singular fact that the great Polynesian name for the Sun-God is also Ra.

[2] *Manners and Customs of the Ancient Egyptians*, Sir J. Gardner Wilkinson.

The Sun-God Ra was usually represented as a man with a hawk's head, surmounted by a globe or disk of the sun, from which an asp issued. His figure, and that of the disk were generally painted in red colour, appropriately suggesting the heat of the mid-day sun. He is sometimes accompanied by a scarabæus or sacred beetle, which was an emblem of the sun throughout Egypt.

Pa-ra, the city of the Sun, or, as the Greeks called it, Heliopolis, was the small but celebrated city of Lower Egypt where Ra was especially worshipped. It lies a little east of the Nile and is not far from the spot where Memphis was built. Its usual name among the Egyptians was An or On. Plutarch has this reference to the Sun worship at Heliopolis: "Those who minister to the god of Heliopolis never carry any wine into the temple, looking upon it as indecent to drink it during the day when under the immediate inspection of their Lord and King."

The priesthood of the Sun were noted for their learning. They excelled in their knowledge of astronomy and all branches of science.

The rat was sacred to Ra, and his votaries were forbidden to eat the rodent.

The best loved of all the Egyptian Sun-Gods, and the first object of their idolatry was Osiris (the one who sees clear), personifying the setting sun. The mysterious and daily disappearance of the orb of day exercised over the Egyptians a phenomenal power. The sun then appeared to them, as Keary beautifully describes it in his *Dawn of History* "to veil his glory and sheathe his dazzling beams in a lovely many-coloured radiance which soothed and gladdened the weary eyes and hearts of men, and enabled them to gaze fearlessly and lovingly on the dreaded orb from which during the day they had been obliged to turn their eyes."

Osiris was distinctly a god of the life eternal. The Egyptians believed that when he sank from sight behind the western hills the souls of the departed were in his retinue, and that, in his nightly sojourn in the underworld, he held high court and judged the dead. Thus, in the inscriptions on the Egyptian temples, we see Osiris in his character as Judge figured in the sacred blue, holding in one hand a sceptre, and in the other the emblem of life, his head surmounted with the double crown of Upper and Lower Egypt.

"In the judgment scenes," says Keary,[1] "he appears seated on a throne at the end of a solemn hall of trial, to which the soul has been arraigned, and in the centre of which stands the fateful balance, where in the presence of the evil accusing spirit and of the friendly funeral gods and genii

[1] *Dawn of History*, C. F. Keary.

who stand around, the heart of man is weighed against a symbol of Divine Truth."

The Sun-God Osiris was therefore to the Egyptians a deity of the living, and a god of the dead, a link connecting the earthly life with the life eternal, the upper and the under worlds, and consequently was personified as two distinct characters. One, as we have seen, depicts him as Judge of the Dead. In his capacity as an earth dweller, Osiris was worshipped under the form of a bull, the Apis, who by successive incarnations never abandoned his home land, and the sight of those who worshipped him through successive ages.

The Egyptians regarded the bull as the living representation of the deity, and believed that the soul of the god tenanted the body of the animal, thence deeming the bull the very same as Osiris himself.

The seat of Osiris worship in Egypt was at Thinis (Teni) in Upper Egypt, on the west bank of the Nile between Thebes and Memphis. Osiris is frequently alluded to as "Lord of Abydos," a city in the immediate vicinity of Thinis, and at this place there have been discovered many temples dedicated to his worship.

That Osiris was a Sun-God is clearly indicated by a number of expressions used regarding him taken from the inscriptions. In the hymns his accession to the throne of his father is compared to the rising of the sun, and it is even said of him in so many words: "He glitters on the horizon, he sends out rays of light from his double feather, inundates the world with it as the sun from out the highest heaven." Like the sun he is called in the sacred songs, "Lord of the length of time." One of his usual appellations is, "Mysterious soul of the Lord of the Disk," or simply "Soul of the Sun."

The Egyptians often called Osiris "Unefer," that is, the good being, representing the beneficent power of the sun that triumphs always over the powers of darkness. In conclusion, the story of the death of Osiris agrees closely with the solar sunset phenomenon, and renders the personification of Osiris as the setting sun a true one.

The most venerable of all the Egyptian Sun-Gods, if not the most popular, was Atum or Amum, personifying the sun after it had set, and was hid from view.

There are two derivations of the name Amum, one meaning, "that which brings to light," the other simply expressing the invitation or greeting of welcome, "Come." In deifying Amum the Egyptians worshipped the unrevealed and unseen Creator of all things, the source of all things, the Ruler of Eternity, whence everything came, and to which all things would return.

On the inscriptions, the figure of Amum was coloured blue, the sacred colour of the source of life; the figure was that of a man with either a human head, or a man's head concealed by the head and horns of a ram. The word "ram" meaning concealment in the Egyptian language.

In the Sun-God Horus we see the dawn personified, and the triumphant conqueror of the shades of darkness and the demons of the underworld emerges in the glorious light of victory each morning. He was figured as the eldest son of Osiris, a strong vigorous youth, who avenged his father by waging a successful war against the monster who had swallowed him up.

Horus is depicted in the inscriptions as sailing forth from the underworld up the eastern sky at dawn, piercing the great python, born of night, as he advances.

"The ultimate victory of life over death, of truth and goodness over falsehood and wrong," says Keary,[1] "were the moral lessons which this parable of the sun's rising read to the ancient Egyptians."

The resemblance of lions to the sun is borne out by the fact that the Egyptians placed the figures of lions under the throne of Horus. This deity was sometimes regarded as the God of Silence,

[1] *Dawn of History*, C. F. Keary.

and represented as a child with his finger held up to his lips.

In addition to the previously mentioned personifications of the Sun, Egyptian Sun worship included a worship of the actual disk of the Sun. This form of worship was in vogue in the reign of Amenophis III., its first appearance on the monuments being in the 11th year of that monarch's reign.

The worship of the solar disk, or Aten, became the sole object of adoration in the reign of Amenhôtop IV. This monarch, in fact, forbade the worship of any god save this, the "great living disk of the sun," and caused the names of all other gods to be erased from the monuments, and their images to be destroyed.

In the hymns, this deity is referred to as he who created "the far heavens and men, beasts and birds; he strengtheneth the eyes with his beams, and when he showeth himself all flowers live and grow, the meadows flourish at his up-going, and are drunken at his feet, all cattle skip on their feet, and the birds that are in the marsh flutter for joy. It is he who bringeth the years, createth the months, maketh the days, calculateth the hours of time by whom men reckon."

In his zeal to make the god of the disk preeminent, King Amenhôtop IV changed his name

to one which signified "gleam of the sun's disk."
The death of this monarch resulted in a great
reaction, the old gods being restored to their
original favour.

Although the solar personifications alluded to
would appear to include all the Sun deities worshipped by the Egyptians, there were several
minor Sun-Gods that had a place in their religion,
chief among these being the god Ptah, personifying
the life-giving power of the sun.   This god was
worshipped with great magnificence at Memphis.

The Sun-God Mandoo personified the power
of the sun's rays at mid-day in summer.   He was
regarded as a god of vengeance and destruction,
and a leader in time of war.

The rays of the sun were personified in the gods
Gom, Moni, and Kons, who are always referred
to as the sons of the Sun-God.   The sun's rays,
personified in the deity Sekhet or Pasht, had a
feminine significance.   This goddess was figured
with the head of a lioness, and it is said she was at
once feared and loved.   Her name Pasht means
the lioness, and was perhaps suggested by the
fierceness of the sun's rays, answering to the lioness's
ferocious strength, or the angry light in her eyes.
Another name for this goddess was "the Lady of
the Cave," and her worship, though common
throughout Egypt, had its seat at Bubastis.

Tiele considers that Set, the enemy and brother of Osiris, was also a Sun-God, as he is sometimes called "the Great Lord of Heaven," and "the Spy." He personified the fierce and terrible desolation wrought by the sun's power.

Chapter VII
Sun Worship (*Continued*)

Chapter VII

Sun Worship (*Continued*)

Sun Worship in India

IN India, a land teeming with mythology, we find as we might expect, Sun worship a predominant feature of the Hindu religion. All the myths prove that the fancied combat between light and darkness, waged daily in the spacious field of the firmament, is of solar origin. As we have seen Osiris, the Sun-God of the Egyptians, triumphing over the demons of darkness, so in India we find Indra, the great solar deity of the Hindus, successful in his combat with Vritra the serpent of night.

The worship of Indra constitutes the very essence of the Vedic religion, although he was by no means the only Sun-God worshipped in India, for the Hindus worshipped the sun in its various aspects after the manner of the Egyptians. The rising sun was called "Brahma," on the meridian it was known as "Siva," and in the west at nightfall, "Vishnu."

"In regard to Vishnu," says Keary, "the great

epic of the Hindus relates that when he was armed for the fight Agni gave him a wheel with a thunderbolt nave. This can only mean a wheel that shoots out thunderbolts from its nave when it turned."[1] The wheel has throughout the ages symbolised the sun.

In Central India, Sun worship still prevails among many of the hill tribes, and the Sun is invoked as the Holy One, the Creator, and Preserver. White animals are sacrificed to him by his votaries.

One of the early and most important Sun-Gods was Sûrya. He is represented as moving daily across the sky in a golden chariot drawn by seven white horses. Perhaps the most holy verse in the Veda is the following short prayer to the Sun-God Sûrya taken from the Rig Veda, an invocation powerful in expression, and beautiful in thought:

"Sing praises unto Sûrya, to the son of Dyaus,
   May this my truthful speech guard me on every side
      wherever heaven and earth and days are spread
      abroad.
   All else that is in motion finds a place of rest. The
      waters ever flow, and ever mounts the sun.
   No godless man from time remotest draws thee down
      when thou art driving forth with wingèd dappled
      steeds.
   He turns him to an alien region of the east, and Sûrya
      thou ariseth with a different light.

[1] *Curiosities of Folk-Lore*, C. F. Keary.

O Sûrya: with the light whereby thou scatterest gloom,
    and with thy ray impellest every moving thing,
Keep far from us all feeble worthless sacrifice, and
    drive away disease and every evil dream.
Sent forth, thou guardest well the path of every man,
    and in thy wonted way ariseth free from wrath.
When, Sûrya, we address our prayers to thee to-day,
    may the gods favour this our purpose and desire.

    .   .   .   .

Ne'er may we suffer want in presence of the sun, and
    living happy lives may we attain our age.
Cheerful in spirit, evermore, and keen of sight, with
    store of children, free from sickness and from sin.

    .   .   .   .

O Sûrya, with the golden hair, ascend for us day after
    day, still bringing purer innocence.
Bless us with shine, bless us with perfect daylight,
    bless us with cold, with fervent heat and lustre.
Bestow on us, O Sûrya, varied riches to bless us in our
    home, and when we travel."

The ancient Sun worship of India is reflected in the daily religious rites and festivals of the modern Hindus. Thus, the time-honoured formula repeated daily since long past ages by every Brahman, indicates clearly the divine element in the sun: "Let us meditate on the desirable light of the divine sun, may he rouse our minds." Here is a direct appeal to a solar deity, and every morning the Brahmans may be seen facing the east, standing on one foot, and stretching out their hands to the sun as they repeat this prayer which has come down unchanged from remote ages.

The Zoroastrians, and the modern exponents of that faith, the Parsees, saw in the sun fire and light, a manifestation of a divine and omnipotent power, and regarded them in a measure as symbols of the deity; but there can be little doubt that this distinction was not always borne out, and that the sun, and fire itself, were literally worshipped by them.

In the Parsee temples burns a fire which, it is said, has never been extinguished since it was kindled by Zoroaster four thousand years ago. In praying, the Parsees are admonished to stand before the fire, and turn their faces toward the sun, and when a young Brahman's head is tonsured, he is to this day so placed that he has the sacred fire to the east whence comes the sun of which it is a type.

Sun Worship in Greece

The ancient Sun-God of the Pelasgians, displaced by the later worship of Apollo, was Arês. "There can be no question," says Keary,[1] "that in prehistoric times the worship of Arês was

[1] *Dawn of History*, C. F. Keary.

widely extended. Traces of his worship are to be found in the Zeus Areios who was honoured at Elis, and in the name of Areiopagus of Athens." Little actual knowledge of this early worship however has come down to us.

The two great solar divinities of Greece were Helios or Hyperion, and Phœbus Apollo. Just as the Egyptians regarded their Sun-Gods Ra and Osiris as distinct aspects of the sun, so the Greeks distinguished the orb from the rays of the sun.

Helios represented to the Greeks the physical phenomenon of light, the orb of the sun which throughout the seasons rises and sets daily. Phœbus Apollo, on the contrary, was the beneficent divinity who not only created the warmth of spring-tide, but protected mankind from the dangers and diseases of the more desolate seasons. He was essentially human in his sympathies and yet wholly godlike in dignity.

Some writers, notably Hesiod, regard Hyperion as the father of the sun, moon, and dawn, and therefore the original Sun-God, and the father of Phœbus Apollo, but Homer identifies Helios with Hyperion as "he who walks on high."

The worship of the Sun-God Helios, the counterpart of the Latin Sol, was imported into Greece from Asia, but by no means gained a high degree of popularity. The number seven was sacred to Helios, and in the island of Trinacria (supposed to be Sicily), it was said he had seven herds of cows, and seven herds of lambs, fifty in each herd, which never increased or diminished in numbers. The god delighted to watch them peacefully grazing when he rose in the morning, and as he left the sky at night-fall. As we shall see later in the chapter on the solar mythology of Greece, and as related in the *Odyssey*, the sacred herds of Helios were ruthlessly slaughtered by the misguided companions of Ulysses. Incensed by this insult the Sun-God threatened to descend into Hades and shine among the dead. He contented himself, however, by complaining to Jove, who, acknowledging the justice of his claim for vengeance, roused up a mighty storm which well-nigh destroyed the miscreants, and completely disabled their ship. Tylor[1] tells us that the Greek Sun-God Helios, to whom horses were sacrificed on the mountain top of Taygetos, was that same personal Sun to whom Socrates, when he had stayed rapt in thought till daybreak, offered a prayer before he departed. An annual festival in honour of Helios was celebrated at Rhodes with musical and athletic contests.

The Greeks believed that the Sun rose out of

[1] *Primitive Culture*, Edward B. Tylor.

the ocean on the eastern side, and drove through the air in a chariot giving light to gods and men. The poet Milton in his *Comus* thus refers to the daily journey of the Sun-God:

"The star that bids the shepherd fold
Now the top of heaven doth hold,
And the gilded car of Day
His glowing axle doth allay
In the steep Atlantic stream,
And the slope Sun his upward beam
Shoots across the dusky pole,
Pacing toward the other goal
Of his chamber in the east."

In Lucian's time the Greeks kissed their hands as an act of worship to the rising Sun.

Shakespeare frequently alludes to Hyperion, and Keats wrote of his downfall, and of the accession of his successor Phœbus Apollo.

We come now to a consideration of the preeminent feature of Hellenic Sun worship, the worship of the Sun-God Phœbus Apollo, the god who is more especially the deity of the later Greeks, the Dorians and Ionians.

The authors of this religion were probably the Dorians, who inhabited the northern portion of Greece, and who founded their first kingdom in Crete. Before the Doric invasion, however, there was in Crete a species of Sun worship, for the bullheaded Minotaur, according to the authorities, could hardly have been anything else than a Sun-God of the Asiatic stamp.

With the coming of the Dorians to Crete, Apollo worship was established, and through the migrations of these people (about the tenth century before our era), the cult of Phœbus Apollo spread on every side, until this religion was in favour wherever the Greek language was spoken.

In Homer, Apollo is easily the greatest of all the Sun-Gods, and superior in character to almost every other deity. In the *Iliad* he is the central and most majestic figure.

Phœbus Apollo, or "Far-Darter" as he is sometimes called, was the son of Jupiter and Latona, and brother of Diana, the Moon Goddess. He was born on Delos, the smallest and most desolate of all the Ægean Islands, after all other places had rejected him. Delos, however, was a most appropriate birthplace for a Sun-God, as the ancients believed that the Sun was born from the sea. His name, Phœbus, signifies the glorious nature of the light of the sun, while the name Apollo probably had reference to the devastating effects of the sun's rays at mid-day.

At the birth of the Sun-God on the seventh day of the month, we are told that sacred swans made the circuit of the island seven times, and all the attendant goddesses gave a shout, and Delos was radiant in golden light. We have here, it is said, the echo of an old belief, that at the hour of sunrise the horizon sends forth a sound.

Zeus bestowed on the infant Apollo a cap, a lyre, and a car drawn by swans. Soon after his birth the swans carried Apollo off to the land of the Hyperboreans, where for six months of the year the climate is marked by sunshine and gentle breezes. Here the Sun-God thrived and waxed vigorous. It is not within the scope of this chapter to dwell on the myths that tell of the mighty deeds of the Far-Darter, as they come properly under the chapter on solar mythology. The establishment, however, of the Delphinian oracle, perhaps the most important event in the life of the Sun-God, is related here, as Delphi was, properly speaking, the seat of Apollo worship.

At an early date Apollo developed the attributes of a warrior, and set out in the quest of adventure. Searching for a suitable place in which to establish an oracle, he came to Delphi, a peaceful vale in Crissa, in the heart of Greece. Its solitude and sublimity completely charmed him and he chose it as the site of his oracle. His advent was not peaceable, however, as Hera had set in his path the great serpent Pytho, and a terrific combat ensued from which Apollo emerged victorious. Some

authorities claim that this contest with Pytho
signifies the war which, according to many mythologies, the Sun-God wages with the River God.
The great river is the earth which flows all around
the world, and which the Greeks knew by the
name "Oceanus."

Perhaps, in the widest significance, this battle
between Phœbus Apollo and the serpent represents the contest between the Sun-God and the
earth river, for the Sun, although seemingly conquered by Oceanus each night, and smothered in
his coils, emerges triumphant in the morning.

From his victory over Pytho, Apollo obtained
the title of "Pythius," and in commemoration of
the event the Pythian games were instituted, in
which contests the victors were crowned with
wreaths of beech leaves.

His foes vanquished, the first requisite of an
oracle, a priest was sought, and it is related that
Apollo cast his eyes seaward and beheld a Cretan
ship sailing for Pylos. Assuming the form of a
dolphin he plunged into the sea, and boarded the
ship to the great amazement of the crew. Under
his guidance they came to the bay of Crissa, and
the god in the form of a blazing star left the ship
and descended into his temple. Assuming the
form of a handsome youth with wavy locks, he
greeted the crew as strangers, and invited them to

The Delphic Sibyl

(Michelangelo)
Sistine Chapel, Vatican

The Ruins of the Greek Temples of Hercules and Apollo, Selinunte,
Sicily

land, and worship him as Apollo Delphinus, as he had met them in the form of a dolphin, and hence Delphi derived its name.

The resemblance between the lives of the Sun-God Phœbus Apollo, and Jesus Christ, the central figure and Exemplar of the Christian religion, is striking. The circumstances of their birth were in many respects similar, in that they were born in comparative obscurity. The mother of Apollo sought in vain for a suitable place to bring forth her offspring, and had recourse at last to a desolate and barren island in the midst of the sea. The Virgin Mary found her only refuge in a comfortless and humble shelter for the beasts of the field. Three gifts were presented the Far-Darter at his birth by Zeus, and the Magi presented the same number of gifts to the infant Jesus. Further, the infant Apollo was hurried away to a peaceful land soon after his birth, and in like manner the child Jesus was conveyed to a place of safety to escape a threatened danger.

For a while Phœbus Apollo hid his greatness in a beggar's garb, bearing with patience the gibes and sneers of his comrades, preferring to bide his time when all men should acknowledge his greatness. This mode of existence was in every way similar to the life of Christ.

Again, as personifying the sun, Phœbus Apollo must necessarily be born weak and suffer hardships, he must wander far and lead a life of strife and action, but above all it was imperative that he should die. It is this last act which makes the character of the Sun-God approach the nearest to human nature.

Although the Sun-God's death at night-fall is ignominious, akin in this respect to the crucifixion, still its predominant feature is one of glory, and the reappearance of the triumphant sun after death is in every way typical of the resurrection, thus portraying in a startling manner the completeness of the analogy between the lives of Christ and Apollo.

In the Homeric hymn to Apollo we read that the Far-Darter took the shape of a dolphin, and guided men from Crete to Crissa. "This plunging of the god into the water, and his taking the shape of a fish," says Keary,[1] "is the setting of the sun, and the birth of Apollo in the mid-Ægean is his rising. Both are alike parts of the sun's daily journey."

Again, the sun is essentially nomadic in its character, a continual wanderer in the firmament and this characteristic is borne out in the life of Phœbus Apollo, who in his infancy started upon his

[1] *Outlines of Primitive Belief*, C. F. Keary.

travels, his life being one of ceaseless wandering and activity.

Apollo brought not only the blessings of the harvest to mankind, but he was the god of music and song. He founded great cities, and promoted colonisation, gave good laws, and in a word, was "the ideal of fair and manly youth, a pure and just god requiring clean hands of those who worshipped him." To him were sacred the wolf, the roe, the mouse, the he-goat, the swan, the dolphin, and the ram. Traces of his solar nature are revealed in some of the statues, which represent him with a full and flowing beard, typifying the sun's rays. Generally he is represented as having the figure of a youthful athlete.

It remains to mention the different titles conferred on Phœbus Apollo. The many festivals inaugurated in his honour are referred to in another chapter. As there is a beneficent side to the sun's character, displayed in its genial warmth, so there is a destructive and desolating force in its rays at mid-day. As a destroyer and producer of plagues Phœbus Apollo was styled "Carneius," and worshipped with particular zeal at Sparta. In this capacity he was also worshipped under the title "Hyacinthus," a worship that was for the most part peculiar to the Peloponnesus.

As a beneficent god the Far-Darter was styled

"Thargelius." The most noted of the temples dedicated to his worship was situated at Amyclæ.

Apollo was regarded as the patron of herdsmen, and in this capacity was called "Nomius." He was also styled "Delphinius," and worshipped in a temple at Athens which bore the name "Delphinian."

From the fact that the number seven was sacred to Apollo he was called "Hebdomeius." As a god of light he was styled "Lycius," the original centre of this worship being Lycia, in the southwest of Asia Minor. Apollo was regarded as the father of Æsculapius, the god of medicine, and those afflicted with disease had recourse to him, for through his kind offices their bodies were purified, and health regained.

In Rome, the worship of Apollo was not established until 320 B.C., a temple being raised to him in that year in consequence of a pestilence that had swept the city. Afterwards a second temple was dedicated to his worship on the Palatine Hill.

To the poets of all ages Phœbus Apollo has been a source of inspiration, and the symbol of poetry. Callimachus, an Alexandrine Greek, who lived about 250 B.C., wrote a hymn in honour of the Sun-God Apollo. In later times, however, no distinction was made by the Greek poets between Apollo and the Sun-God Helios.

The poet Keats wrote an ode and a hymn to
Apollo, and Shelley's *Hymn to Apollo* is considered one of the finest and most sublime poems
in our language. It is the Sun-God's description
of his divine attributes, and because of its beauty
is quoted in full:

"The sleepless Hours who watch me as I lie,
Curtained with star-inwoven tapestries
From the broad moonlight of the sky,
Fanning the busy dreams from my dim eyes.
Waken me when their mother, the grey Dawn,
Tells them that dreams and that the moon is gone.

"Then I arise, and climbing Heaven's blue dome,
I walk over the mountains and the waves,
Leaving my robe upon the ocean foam;
My footsteps pave the clouds with fire; the caves
Are filled with my bright presence, and the air
Leaves the green earth to my embraces bare.

"The sunbeams are my shafts with which I kill
Deceit, that loves the night and fears the day;
All men who do or even imagine ill
Fly me, and from the glory of my ray
Good minds and open actions take new might
Until diminished by the reign of night.

"I feed the clouds, the rainbows, and the flowers
With their ethereal colours; the moon's globe
And the pure stars in their eternal bowers
Are cinctured with my power as with a robe;
Whatever lamps on Earth or Heaven may shine
Are portions of one power, which is mine.

"I stand at noon upon the peak of Heaven,
  Then with unwilling steps I wander down
  Into the clouds of the Atlantic even;
  For grief that I depart they weep and frown.
  What look is more delightful than the smile
  With which I soothe them from the western isle?

"I am the eye with which the Universe
  Beholds itself and knows itself divine;
  All harmony of instrument or verse,
  All prophecy, all medicine are mine,
  All light of Art or Nature; to my song
  Victory and praise in their own right belong."

The Dorians, as they migrated and founded new kingdoms, found the worship of the Sun-God Herakles flourishing in other lands, and gradually this form of religion became popular and supplanted the worship of Apollo.

Herakles was, however, considered more in the light of a solar hero than a Sun-God. In Herakles we behold the Sun, loving and beloved, wherever he goes seeking to benefit the sons of men; yet, as was the case with Apollo, sometimes bringing destruction and desolation down upon them through the fierce heat of his noonday rays.

The twelve labours of Herakles are supposed to refer to the sun's passage through the Zodiacal signs, as they suggest forcibly in many cases the successive conquests of the Sun hero. Herakles was represented on coins of Cyzicus about 500–

450 B.C. The death of Herakles is the most impressive incident of his varied career. "No one," says Keary,[1] "who reads the account of it, can fail to be struck by the likeness of the picture to an image of the setting sun. . . . The flame of his pyre shines out far over the sea, and the sun's last rays shine out in the light of the fiery sky." The many myths related of the Sun hero, Herakles, are referred to in the chapter on solar mythology.

Japanese Sun Worship

In the spirit religion of Japan we find the worship of the Sun-God is supreme. He is regarded as the "heaven-enlightening great spirit,—below him stand all the lesser spirits through whom as mediators, guardians, and protectors, worship is paid by men."

Among the Shinto deities, however, the Sun-Goddess was the central figure. To reconcile Buddhism and Shintoism the chief priests claimed that the Sun-Goddess had been merely an incarnation of Buddha.

The shrine of the Sun-Goddess stood in the Mikado's residence, and was reverenced by that monarch as one of his family gods. Her emblem was the mirror, which is to the present day con-

[1] *Dawn of History*, C. F. Keary.

sidered one of the sacred treasures of the Japanese sovereigns.

A temple of the Sun-Goddess was established at Watarahi in the province of Isé, and the shrine of the Goddess of Food was placed in the temple. These two deities henceforth occupy together the chief place in the Japanese Pantheon. They were honoured above all other gods by festivals and ceremonies held annually. Offerings and sacrifices were presented to these goddesses on the seventeenth day of the sixth moon, and the ritual of the invocation was in part as follows: "Hear, all you ministers of the gods, and sanctifiers of offerings, the great ritual declared in the presence of the From-Heaven-Shining-Great Deity."

At the harvest festival thanksgiving was offered to the Sun-Goddess for bestowing upon her descendants dominion over land and sea.

Sun Worship in Peru and Mexico

From Japan we cross the Pacific to find the indigenous tribes of the western continents reverencing and worshipping the Sun in ancient times. The Sun worship of Peru first claims our attention, as it easily overshadows in importance and magnificence the solar worship of any other of the western tribes.

It has been shown that Sun worship prevails, for the most part, where the sun is welcomed for his genial warmth, and where nature suffers at his departure. Thus, in the lowlands of South America, Sun worship attained little prominence, but on the high plateaus, such as those in Peru, it flourished vigorously, and was the dominant feature of the life of the natives.

The Peruvians believed that the Sun was at once the ancestor and the founder of the Inca dynasty, and that the Incas reigned as his representatives and almost in his person. The Sun, therefore, was the sovereign lord of the world, the king of heaven and earth, and was called by them "Inti," which signifies Light.

The Peruvian villages were so built that the inhabitants could have an unobstructed view of the east, in order that each morning the nation might unite in saluting the rising Sun, and rejoice in the advent of the Lord of Light. The Sun alone of all the deities had a temple in every large town in Peru.

The Peruvian Sun temples probably exceeded in magnificence those of any other nation on the earth. In Peru, as elsewhere, a certain relationship was thought to exist between the substance of gold, and that of the sun. In the nuggets dislodged from the mountain sides they thought they saw the Sun's tears, consequently, in the Peruvian edifices dedicated to the worship of the Sun, we find gold used lavishly to beautify and embellish the structure.

The following description of the Great Temple of the Sun at Cuzco, the grandest ecclesiastical edifice in the empire, gives some idea of the beauty and grandeur of these places of worship:

The temple consisted of a vast central auditorium surrounded by a number of smaller buildings and was built with an elegance of masonry rarely, if ever, equalled.

The roof was formed by timberworks of precious woods plated with gold, and the precious metal was so prodigally lavished on the interior that the temple bore the name of "The Place of Gold" or "Golden Palace." A thick sheet of gold six inches wide ran round the outside of the edifice as a frieze, and there was a similar decoration in every apartment. The doors opened to the east, and at the far end above the altar was a golden disk with human countenance shaped and graven to represent the sun, and studded with precious stones. It was so placed as to reflect, at certain seasons, the first rays of the rising sun on its brilliant surface, and, as it were, reproduce the likeness of the great luminary.

Around the sacred disk was arranged in a semi-

The Ruins of the Temple of the Sun, Cuzco, Peru

Courtesy of Mr. Leon Campbell

The Ruins of the Temple of the Sun, Cuzco, Peru

Courtesy of Mr. Leon Campbell

circle the mummies of the departed Incas seated
on golden thrones, so that the morning sun rays
came day by day to bless the remains of the rulers
of bygone ages.

The adjacent buildings were the abodes of the
deities who formed the retinue of the Sun. The
principal one was sacred to the Moon, the Sun's
consort, who had her disk of silver, and arranged
around her were the mummies of the ancient
queens. Other chambers in the temple were
dedicated to the stars, to lightning, and to the
rainbow. Outside the temple was a great garden
filled with rare and beautiful plants, which contained, also, exquisite imitations of trees, bushes,
and flowering shrubs, and animals all wrought in
solid gold. The vases and temple ornaments,
all the utensils used by the priests in the temple,
and even the conductor pipes, were composed of
the precious metal.

In the Peruvian ceremonials of Sun worship,
drink offerings were presented to the deity in a
golden vessel, and the people believed that if the
liquid disappeared the Sun partook of it, which
might be truly said of it, as it soon vanished by
evaporation.

Under the Incas, Sun worship became the state
religion of Peru, and the central idea of the life
of the people. It is evident, however, that Sun

worship was not acceptable to all the Incas, for there is on record a protest made by an Inca that the Sun could not be a supreme and all-powerful deity, constrained as he was to pursue one fixed course in the firmament. If he was supreme he should be a free agent, argued this wise sovereign.

Columns in honour of the Sun were erected in Peru as in other lands where Sun worship prevailed, level at the top, so as to form a seat for the sun who, the Peruvians said, "loved to rest upon them." At the equinoxes and solstices they placed golden thrones upon them for the Sun-God's further convenience. Surrounding the city of Cuzco there were twelve stone columns dedicated to the sun, which represented the twelve months in the year.

Human sacrifices to the sun were common in Peru, and the rising sun looked down on sacrificial altars reddened by the blood of thousands of victims. The holiest sacrifice was the blood of a captive youth, smeared on a rock that crowned a mountain top, so that the sun's first rays would light up the gory sacrifice.

Sun worship thrived in Peru until the Spanish Conquest, when Pizarro ruthlessly overthrew the temples, and stifled the religion. It is said that the great golden disk representing the sun, that was the chief object of worship in the Great Temple at Cuzco, was secured as booty by one of the rough adventurers of the conquering army, and became the stake in a subsequent drunken gambling bout that the plunderer engaged in.

Although the Sun worship of the Peruvians reached a higher state of exaltation, and perfection than that of any other South American people, still the pre-eminence of the Sun, and its deification, was the very essence of the early religion of Central America, and particularly Mexico. The ancient Mexicans called themselves "Children of the Sun," and daily greeted the rising sun with hymns of praise, and offered to the solar deity a share of their meat and drink. Even to this day, the inhabitants of the interior of Mexico, as they go to mass, throw a kiss to the Sun before entering the church.

Four times by day and night the priests of the ancient Sun temples addressed their invocations and prayers to the Sun, and all the temples were dedicated to his worship. In the ceremonial of the temple worship, blood drawn from the ears of the high priest was offered to the Sun, as was also a sacrifice of quails. The priest invoked the Sun saying: "The Sun has risen, we know not how he will fulfil his course, nor whether misfortune will happen. Our Lord do your office prosperously."

The temples of the ancient Toltecs, who inhabited Mexico as far back as the year 674 of the

Christian era, were dedicated to the Sun. The Moon they worshipped as his wife, and the Stars as his sisters. No image was allowed within these temples, and their offerings were perfumed flowers, and sweet-scented gums. They reared in adoration of the Sun and Moon great pyramids which have endured to this day and examples of which may be seen at San Juan, Teotihuacan.

The highest "El Sol" is 216 feet in height, has a base about 761 feet square, and the summit is reached by a flight of sixty-eight steps. Many strange idols have been found in this region embellished and ornamented with designs of solar significance, the sun's rays being especially noticeable in the carvings.

The supreme god of the early Mexicans was Quetzalcohuatl, who personified to them the Sun of to-day; his father was Camaxtli, the great Toltec conqueror, whom the Mexicans regarded as the Sun of yesterday:—while the god Tetzcalipoca signified to them the Sun of to-morrow.

Quetzalcohuatl is described as being regal of stature, of white complexion, and of pleasing countenance. His face was fair, and his beard bushy, and he was clothed in flowing robes. It is related that the clouds, or cloud snakes, bear down the old Sun, and choke him, but the young Quetzalcohuatl rushes up in the midst of them from

Pyramid of the Sun, San Juan, Teotihuacan

Permission of Detroit Publishing Co.

Stonehenge, England

below, and destroys them. As we shall see, the Sun-God Hercules, the solar deity of the Greeks, was famed even in his infancy for his triumph over the serpents sent to destroy him.

Quetzalcohuatl reigned over the Toltecs peaceably for many years, but finally his enemies brought about his downfall, and deposed him. Legend says he embarked in his ship and sailed down a river to the sea, where he disappeared and was no more seen. When Montezuma beheld Cortez and the Spanish ships approaching the land, he thought that the great Sun-God was returning to his beloved land.

Sun Worship of the North American Indians

Proceeding northward, we find the worship of the Sun that anciently existed and flourished in the far east, equally prominent in the life of the early Indian tribes of North America.

The chiefs of the Huron tribe claimed descent from the Sun, and believed that the sacred pipe was derived from this luminary. It was, they thought, first presented to the Pawnees, and by them transmitted to the other tribes. Many of the Indian tribes have a similar tradition.

The Iroquois regarded the Sun as a god, and offered him tobacco, which they termed "smoking the Sun." On important occasions the braves gathered together in a circle, squatting on the ground, the chief then lighting the calumet, and offering it thrice to the rising Sun, imploring his protection, and recommending the tribe to his care. The chief next took several puffs and passed the pipe on for all the others to smoke in turn. As in many Christian churches to-day, the prayers of the people mingle with the smoke of incense, so were the invocations of the early Indians addressed to their Sun deity, supposed to be wafted to him by the smoke that wreathed upward from the sacred calumet.

Certain tribes offered to the Sun the first game they despatched when they were out on a hunting expedition. The Apalachees of Florida, in their sacrifices to the Sun, offered nothing that had life. They regarded the Sun as the parent of life, and thought that he looked with displeasure on the destruction of any living creature. They saluted the Sun at the doors of their wigwams as he rose and set, and in the sacred hut or cave where they worshipped, the Sun's rays were permitted to enter so as to illuminate the altar at certain times of ceremonial importance. This accords closely with the ideas of orientation which played such an important part in the Sun temple worship of the Egyptians. In the course of their service of Sun worship, the Apalachees released the sacred Sun birds through a crevice in the roof of the cave temple. These, as they winged their way upward, were thought to convey their expressions of adoration to the Sun, the supreme deity.

To the Creeks the Sun represented the Great Spirit, toward him they directed the first puff of smoke from the calumet, as they sat in solemn council, and to him they bowed reverently in their discussions. The early Indian tribes of Virginia prostrated themselves before the rising and setting Sun and Tylor[1] tells us that the Pottawotomies would climb sometimes at sunrise to the roofs of their huts to kneel, and offer to the luminary a mess of Indian corn.

The powerful Sioux tribe regarded the Sun as the Creator and Preserver of all things and to him they sacrificed the best of the game they killed in the hunt. The Shawnees believed that the Sun animated everything, and therefore must be the Master of Life or Great Spirit.

The Sun worship of the Indian tribes dwelling in the southern portions of North America seems to have been on a more elaborate scale than that in vogue in the north. Doubtless it was influenced by the widely extended and exalted Sun worship of the South American tribes. Among the tribes inhabiting what is now the state of Louisiana, it

[1] *Primitive Culture*, Edward B. Tylor.

was customary for the chief to face the east each morning, and prostrate himself before the rising Sun. He also smoked toward it, and then toward the three other cardinal points of the compass. These Indians even erected to the Sun a rude temple, a circular hut some thirty feet in diameter. In the midst of it was kept burning a perpetual fire, prayers were offered to the Sun three times each day, and the hut was the repository of images and religious relics. Following the Inca custom, the bones of their departed chiefs were also placed in the sacred structure. Their highest and most powerful chief was regarded as the Sun's brother, and he conducted as high priest the temple service of worship to the Sun.

The Dakota Indians called the Sun "the mysterious one of day," and believed that this deity watched over them in time of need. The following translation from portions of an Indian hymn to the Sun indicates the attitude of the worshippers toward their deity:

Great Spirit, master of our lives,
Great Spirit master of all things visible and invisible,
    and who daily makes them visible and invisible,
Great Spirit master of every other Spirit good or bad,
Command the good to be favourable unto us, and
    deter the bad from the commission of evil.

.   .   .   .

O Great Spirit, when hidden in the west protect us
from our enemies who violate the night, and do
evil when thou art not present,
Make known to us your pleasure by sending to us the
Spirit of Dreams.

. . . .

O Great Spirit, sleep not longer in the gloomy west, but
return and call your people to light and life.

Brinton[1] tells us that the Algonquins did not
regard the Sun as a divinity but merely as a symbol. They called the Sun "Wigwam of the Great
Spirit," and they prayed not to the Sun but to the
old man who dwelt in the Sun.

Fire worship is closely related to Sun worship,
and in many cases the North American Indians
regarded fire, and not the Sun, as the Creator of
all things, and the Supreme Deity.

The Choctaws refer to fire as "the greater chief,"
and speak of it as "he who accompanies the Sun,
and the Sun him." In preparing for war they
invoke the aid of both the Sun and Fire.

The following Ottawa legend[2] is of much interest
as showing clearly the motives with which savage
animists offer sacrifices to their deities, and the
spirit in which they believe the gods accept them:

"Onowuttokwutto, the Ojibwa youth who has
followed the moon up to the lovely heaven prairies,

[1] *Myths of the New World*, Daniel G. Brinton.
[2] *Primitive Culture*, Edward B. Tylor.

to be her husband, is taken one day by her brother the Sun to see how he gets his dinner. The two look down together through the hole in the sky upon the earth below. The Sun points out a group of children playing beside a lodge, at the same time throwing a tiny stone to hit a beautiful boy. The child falls, they see him carried into the lodge, they hear the sound of the rattle, and the song and prayers of the medicine man that the child's life might be spared. To the entreaty of the medicine man the Sun makes answer: 'Send me up the white dog.' Then the two spectators above could distinguish on the earth the hurry and bustle of preparation for a feast, a white dog killed and singed, and the people who were called assembling at the lodge. While these things were passing the Sun addressed himself to his youthful companion saying:—'There are among you in the lower world some whom you call great medicine men, but it is because their ears are open, and they hear my voice, when I have struck any one, that they are able to give relief to the sick. They direct the people to send me whatever I call for, and when they have sent it I remove my hand from those I have made sick.' When he had said this the white dog was parcelled out in dishes for those that were at the feast, then the medicine man, when they were about to begin to eat said, 'We send thee this, Great Manito.' Immediately the Sun and his companion saw the dog, cooked and ready to be eaten, rising to them through the air, and then and there they dined upon it."

Of all the Indian customs and forms of worship of solar significance, the great ceremonial of the Sun dance best exemplified their worship of the Sun. Although it partook of the nature of a solar festival, and might properly be included in the chapter on Solar Festivals, still the deep religious significance of this rite, and the fact that it was essentially an act of Sun worship, renders it in accord with the subject under discussion. The following description of the Sun dance of the Senecas is taken from volume xxiii., of the *Journal of American Folk-Lore:*

"The Seneca sun dance is called by any individual who dreams that the rite is necessary for the welfare of the community. It begins promptly at high noon, when three showers of arrows or volleys from muskets are shot heavenward to notify the sun of the intention to address him. After each of the volleys the populace shout their war cries 'for the sun loves war.' A ceremonial fire is then built, and the sun priest chants his thanksgiving song, casting from a husk basket handfuls of native tobacco upon the flames as he sings. This ceremony takes place outside of the

Long House where the rising smoke may lift the words of the speaker to the sun. Immediately after this the entire assemblage enters the Long House where the costumed feather dancers start the Ostowā"gowa. Among the Onondaga of the Grand River Reserve in Ontario the leader of the sun ceremony carries an effigy of the sun. This is a disk of wood ten inches in diameter, fastened to a handle perhaps a foot long. The disk is painted red, and has a border of yellow. Around the edge are stuck yellow-tipped down feathers from some large bird to represent the sun's rays."

The great tribal ceremony of the Kiowas was the Sun dance, which was generally celebrated each year about the middle of June. It lasted four days, and during this time the sacred image representing a human figure, and supposed to possess magical qualities, was exposed in the medicine lodge. It was the only time in the year that the sacred image was revealed to the people, and the veneration of the object was similar in many respects to the worship paid to the alleged relics of the saints that are now extant.

The Kiowas considered even the accidental shedding of blood at a Sun dance an evil omen, and their ceremonies were free from the horrible acts of self-torture that made the Sun dances of many of the Indian tribes especially revolting.

A description of the Sun dance of the Sioux follows.[1] This is given in much detail as the subject is of prime importance in any discussion of the solar ceremonials of the Indian tribes.

After the day for the Sun dance had been appointed by the medicine men, a straight and tapering pine, forty or fifty feet high, was selected for the sun pole. This was chosen by the oldest woman in the camp, and the task of stripping it of boughs and foliage, and clearing a passage about it, was left to the gaily dressed maidens. This work was performed on the second day of the ceremony. Before sunrise the next day a long line of naked young warriors was formed, gorgeous in war paint and feathers, bearing their weapons. This line was drawn up facing the east, and the sun pole which was five or six hundred yards away.

Overlooking the scene, on a high hill, stood an old medicine man, whose sole duty was to signal the moment of the sun's rising. Suddenly the signal was given, and with a great shout the Indians mounted their ponies and rode straight at the pole, discharging their weapons at it as they advanced. Chips from the pole flew in all directions, and if it fell a new pole must be selected. Later in the day the pole was cut down, and set up in the centre of

[1] F. Schwatka in vol. xvii., *Century Magazine.*

a great plain, guy ropes of buffalo thongs, diverging from its top, being used to steady it. These ropes were fastened at the lower end to the tops of stakes which were driven in the ground at regular intervals about the sun pole, forming a circle about it.

Early on the morning of the third day the true worship of the Sun was begun, and a number of young warriors who had prepared for the ordeal by fasting for a number of days, presented themselves, being placed facing the sun. They were arrayed in full war paint and feathers, with fists clenched across their breasts; jumping up and down in measured leaps they circled about, keeping time to the monotonous beating of the tom-toms. Now and then a similar group of young maidens would appear in another part of the arena, and take up a song. The dancing continued for intervals of from ten minutes to a quarter of an hour, broken by rests of about equal length, and lasted from sunrise to sunset. During the day horses were brought into the arena, and the medicine men, after many incantations, dipped their hands into coloured earth, and smeared it on the flanks of the animals.

On the fourth day of the Sun dance the self-torture began, the male dancers of the previous day participating in this rite. The row of dancers took their places promptly at sunrise, but it was not before nine or ten o'clock that the torture began. Each of the young men presented himself to a medicine man who took between his thumb and forefinger a fold of loose skin of the breast, and a skewer, which had been previously fastened to the lower end of a guy rope which supported the sun pole, was then passed through the victim's flesh. The object of the devotee was to break loose from this fetter which bound him to the sun pole without using his hands. This could only be accomplished by so straining against his bonds that the skewer was torn free from the skin through which it passed. The torture was frightful, and frequently the victims fainted under the ordeal. All the while the beating of the tom-toms and the weird chanting of the singers continued. When the day was about over the survivors of the ordeal of torture filed from the arena one by one, and just outside of it they knelt with arms crossed over their bleeding breasts, and with bowed heads faced the setting sun. They rose only when it had disappeared from view.

It remains to refer briefly to evidences of Sun worship in various parts of the world, and the survivals of this cultus in the religious observances and ceremonials of to-day.

In Rome, in the fifth century, it was the custom to bow to the sun before entering a church, and to salute the rising sun from the summit of a hill.

The Emperor Constantine was an ardent votary of the sun, and it was a great triumph for Christianity when he forsook this form of idolatry.

In Armenia forms of Sun worship exist to-day, and the Bedouins of the Arabian Desert constantly practise the adoration of the rising Sun, in spite of the Prophet's command against such observances.

In the Upper Palatinate it is the custom to take off the hat to the rising sun, and in Pomerania the fever-stricken patient is admonished to face the rising sun, and invoke the sun thrice as follows: "Dear Sun, come down soon, and take the seventy-seven fevers from me, in the name of the Holy Trinity."

The rude Tartar tribes sacrificed their horses to the Sun-God, whom they say frees them from the miseries of winter, and Mongol hordes may still be met with whose high priest prays to the sun, and throws milk up into the air as an offering to the Sun-God.

In Australia and Polynesia solar mythology overshadows the deification of the sun, and the early history of these regions is rich in legends and tales of the mighty deeds of the sun hero.

No chapter on Sun worship would be complete without some reference to Stonehenge, for Druidical worship embraced, even if it was not entirely governed by, the sacred rites of solar worship.

The position and location of the group of stones on the wide plain commanding an unobstructed view of the horizon, reveal the character of the worship of those who placed them there as solar in its nature. Those who annually meet at Stonehenge in the present day, at dawn on the twenty-first of June, note that the sun rises exactly over the centre of the stone known as "the Pointer." It has been thought by some students that the Sun worship of the Druids was introduced into England and Ireland by Egyptian colonists, as the rites of the Druids conform in a remarkable degree with those attached to Sun worship in ancient Egypt. A list of the Sun-Gods of the various nations follows, with the authorities that establish their respective claims to solar deification.[1]

Deities Declared to be the Sun	Authorities
Saturn or Cronus	Macrobius, Nonnus
Jupiter	" "
Pluto or Aidoneus	The Orphic Poet
Bacchus or Dionysus	Virgil, Ausonius, Macrobius, Sophocles
Priapus	The Orphic Poet
Apollo	All authorities
Janus	Macrobius

[1] *The Origin of Pagan Idolatry*, G. S. Faber.

Deities Declared to be the Sun	Authorities
Pan	Macrobius
Hercules	" Nonnus
Vulcan	The Orphic Poet
Æsculapius	Macrobius
Mercury or Hermes	"
Osiris, Horus, Serapis	Diodorus Siculus, Macrobius, Eusebius.
Belus or Baal	Nonnus
Adonis	Macrobius

The following list indicates the principal titles given to the Sun-Gods by the ancient nations.

Title	Nation
Baal or Belus	Chaldeans, Assyrians, Moabites
Shamash	Babylonians
Moloch	Ammonites
Moloch, Baal, Chemosh, Baal-Zebub, Thammuz	Canaanites
Hammon	Libyans
Saturn	Carthaginians
Adonis, Baal, Melkarth, Bel-Samen	Phœnicians
Asabinus	Ethiopians

Title	Nation
Ra, Osiris, Horus, Atum, Ptah, Mandoo, Gom, Moni, Kons, Sekhet, Pasht, Set	Egyptians
Mithras	Persians
Brahma, Vishnu, Siva, Indra, Sûrya	Hindus
Adonis, Dionysus, Urotal	Arabians
Belenus	Gauls
Arês, Herakles, Apollo, Bacchus, Liber, Dionysus	Greeks, Romans
Hu	Druids
Viracocha	Peruvians
Vitzliputzli	Mexicans

In conclusion, no claim is made that the subject of Sun worship has been treated exhaustively in the foregoing fragmentary account of this ancient form of idolatry in many lands. The main purpose has been to indicate how widespread solar deification was, and how, at one time or another, it has been the central and predominant feature of the religious life of all people.

Again, the similarity of the ceremonials and forms of Sun worship, the sacrifices and rituals of the various and widely separated nations of the earth, provide features that cannot fail to interest the student of history and ethnology.

Some of these points of resemblance are so striking as to suggest that at an early date in the world's history there was some means of communication between, or link that joined, the eastern and western continents.

Because of the far-reaching influences of Sun worship—influences that exist to-day—and by reason of its importance in the life and history of the ancient nations, the subject must ever remain one of the most interesting and absorbing that the life pages of the race record.

Chapter VIII
Sun-Catcher Myths

Chapter VIII

Sun-Catcher Myths

IN the mythology of every land there may be found legends relating to the snaring of the sun, or the retarding of its daily course. The regularity and the steadfastness of the sun's apparent diurnal motion, and its undeviating and deliberate journey across the sky, was such a peculiar and obvious circumstance that it was a matter of conjecture and speculation at a very early date in the world's history. The cause of this phenomenon was naturally attributed to compulsion; for why should the sun mount the sky each morning with absolute regularity, and pursue the same path always unless compelled to do so?

Some time or other the Sun probably did as he liked, and doubtless he was caught in a trap, and beaten into submission; or, perhaps formerly he went rapidly across the heavens, but, being caught he was forced to proceed at a more leisurely gait. Thus the ancients speculated regarding the daily apparent movement of the sun, and these notions gave rise to a wealth of tradition, myth, and legend that have come down to us in many devious ways. The tales that relate especially to the snaring and trapping of the Sun have been termed by mythologists "Sun-Catcher Myths."

In a pass of the Andes there stand, on the cliffs that rise high on either side, two ruined towers. Into their walls are clamped iron hooks, which tradition relates held fast a net that was stretched across the pass to catch the rising Sun. According to an Indian legend, the Sun was once caught and bound with a chain which only permitted him to swing a little way to one side or the other.

It is said that Jerome of Prague, when travelling among the heathen Lithuanians, early in the fifteenth century, found a tribe who worshipped the Sun, and idolised a large iron hammer. The priests informed him that once the Sun had been invisible for many months because a powerful king had imprisoned him in a strong tower, but the signs of the zodiac had broken open the tower with this very hammer, and released the Sun, therefore they worshipped it.

A Japanese myth relates that in early times the Sun, displeased at men's misdeeds, retired into a cave, and left the world in darkness.[1]　This caused

---

[1] Here again we encounter in a land far distant the tradition common among the American Indians, that at one time the land they dwelt in was shrouded in darkness.

great distress, but finally the wise men devised a plan to lure her from her retreat, which plan was successful. When the Sun discovered the ruse, she desired to forsake the world once more, but, before she could do so, she was bound by cords, and held fast by eight hundred thousand gods, who have ever since restrained her from leaving the world.

As we have seen, the worship of the Sun was abandoned for a time in Peru, as one Inca denied that the Sun was a supreme deity because he followed a circumscribed course. "If he were free," said the Inca, "he would visit other parts of the heavens where he had never been. As he follows one path, he must be tied like a beast who goes ever round and round in the same track."

Both in the Orient and Occident we find myths relating to the subjection of the Sun, but whereas the culture of the East invented the beautifully adorned legends of Phœbus Apollo, and the mighty Herakles, who, although all-powerful, was doomed to a life of servitude at the behest of another, in the West the crude imagination of the barbarians conjured up the mental picture of the snaring of the Sun by an artfully contrived net or trap.

The association of the Sun with cords or ropes in the myths is clearly derived from the phenomenon in evidence when the sun's rays filter through a broken mass of clouds; then the long streams of light that seem to radiate from the sun and touch the earth are in imagination not unlike strands of gold which, as the legend related, held the Sun fast bound to earth.   Thus did they appear to the Polynesians, who call these rays of the sun the ropes by which the Sun is fastened. They say the Sun once drove swiftly through the sky, but a god subdued him, and now bound by ropes and cords he goes humbly along his daily appointed path.

The Polynesian myth[1] that tells of this snaring of the Sun is one of the most interesting legends in solar mythology, and it is therefore given in much detail:

Maui, the Polynesian hero god, after performing many great exploits, returned home to dwell with his brothers.   He soon became restless, however, and, looking about for adventure, he decided that the Sun's daily course across the sky was altogether too rapid, and night-fall followed dawn too quickly to suit him.

"So at last one day he said to his brothers: 'Let us now catch the sun in a noose, so that we may compel him to move more slowly, in order that mankind may have long days to labour in to procure subsistence for themselves.'   But they

[1] *Polynesian Mythology*, Sir George Gray.

answered him: 'Why, no man could approach it on account of its warmth and the fierceness of its heat.' But nothing daunted, Maui persisted that he could snare the sun, and with the aid of his brothers spun and twisted ropes to form a noose. After they had sufficient ropes with which to bind the sun, Maui and his brothers journeyed a long distance eastward to the very edge of the place out of which the sun rises. Then they set to work, and built on each side of this place a long high wall of clay with huts of boughs of trees at each end to hide themselves in. When these were finished, they made the loops of the noose, and the brothers of Maui then lay in wait on one side of the place out of which the sun rises, and Maui himself lay in wait upon the other side.

"The young hero held in his hand his enchanted weapon, the jaw bone of his ancestress, and said to his brothers: 'Mind now, keep yourselves hid, and do not go showing yourselves foolishly to the sun, if you do you will frighten him; but wait patiently until his head and forelegs have got well into the snare, then I will shout out, haul away as hard as you can on the ropes on both sides, and then I'll rush out and attack him, but do you keep your ropes tight for a good long time (while I attack him), until he is nearly dead, when we will let him go, but mind now, my brothers, do not let him move you to pity with his shrieks and screams.'

"At last the sun came rising out of his place like a fire spreading far and wide over the mountains and forests. He rises up, his head passes through the noose, and it takes in more and more of his body until his fore-paws pass through, then are pulled tight the ropes, and the monster began to struggle and roll himself about, whilst the snare jerked backwards and forwards as he struggled. Then forth rushed that bold hero Maui with his enchanted weapon. Alas, the sun screams aloud, he roars; Maui strikes him fiercely with many blows; they hold him for a long time. At last they let him go, and then weak from wounds the sun crept slowly along its course. Then was learnt by men the second name of the sun, for in its agony the sun screamed out: 'Why am I thus smitten by you? Oh man: Do you know what you are doing? Why should you wish to kill Tama-nui-te-ra?' Thus was learnt the second name. At last they let him go, and the sun went very slowly and feebly on his course.

"Maui, however, took the precaution to keep the ropes on him, and they may still be seen hanging from the sun at dawn and eve."

It is also related that in snaring the Sun, Maui injured it, and thus deprived it of half its light, and since then the days have been longer and cooler and men have been able to work in peace.

The following Sun-catcher myth[1] refers more to a temporary staying of the Sun in its daily course, than to a permanent change in its rate of speed such as Maui effected:

"There was once a man who, like the white people, though it was years before pipes, muskets, or priests were heard of, never could be contented with what he had. Pudding was not good enough for him, and he worried his family out of all heart with his new ways and ideas. At last he set to build himself a house of great stones to last forever. So he rose early and toiled late, but the stones were so heavy, and so far off, and the sun went around so quickly, that he could get on but very slowly. One evening he lay awake, and thought, and thought, and it struck him that as the sun had but one road to come by, he might stop him, and keep him till the work was done. So he rose before the dawn, and pulling out in his canoe, as the sun rose, he threw a rope around his neck, but no, the sun marched on, and went his course unchecked. He then put nets over the place where the sun rose, he used up all his mats to stop him, but in vain, the sun went on, and laughed in the hot winds at all his efforts. Meanwhile the house stood still,

[1] *Researches in the Early History of Mankind*, Edward B. Tylor.

and the builder fairly despaired.   At last the great
Itu, who generally lies on his mats, and cares not
at all for those he has made, turned round and
heard his cry, and because he was a good warrior
sent him help.   He made the facehere creeper grow,
and again the poor man sprang up from the ground
near his house, where he had lain down in de-
spair.   He took his canoe, and made a noose of the
creeper.   It was a bad season when the sun is dull,
and heavy, so up he came half asleep and tired,
nor looked about him, but put his head into the
noose.   He pulled and jerked, but Itu had made it
too strong.   The man built his house, the sun
cried and cried till the island of Savai was nearly
drowned but not till the last stone was laid was he
suffered to resume his career.   None can break
the facehere creeper.   It is the Itu's cord."

A study of North American mythology reveals
in the traditions of the Ojibways a myth similar
in many respects to the Polynesian Sun-catcher
myth.   This legend[1] relates that in primitive times
the animals ruled the earth, having killed all of
humankind except a girl and her small brother.
They lived in fear and seclusion:

"The boy never grew bigger than a little child,
and his sister used to take him out with her when
she went to get food for the lodge-fire, for he was

[1] *Researches in the Early History of Mankind*, Edward B. Tylor.

too little to leave alone. A big bird might have flown away with him. One day she made him a bow and arrows, and told him to hide where she had been chopping, and when the snow birds came to pick the worms out of the wood, he was to shoot one. That day he tried in vain to kill one, but the next, toward night-fall, she heard his little footsteps on the snow. He brought in a bird, and told his sister she was to take off the skin, and put half the bird at a time into the pottage, for till then men had not begun to eat animal food, but had lived on vegetables alone. At last the boy had killed ten birds, and his sister made him a little coat of the skins. 'Sister,' said he one day, 'are we all alone in the world? Is there nobody else living?' Then she told him that those they feared, and who had destroyed their relatives, lived in a certain part, and he must by no means go that way, but this only made him more eager to go, and he took his bow and arrow and started. When he had walked a long while, he lay down on a knoll, where the sun had melted the snow, and fell fast asleep, but while he was sleeping the sun beat so hot upon him, that his bird-skin coat was all singed and shrunk. When he awoke and found his coat spoilt, he vowed vengeance against the sun, and bade his sister make him a snare. She made him one of deer's sinew, and then one of her own hair, but they would not do. At last she brought him one that was right. He pulled it between his lips, and as he pulled it became a red metal cord. With this he set out a little after midnight, and fixed his snare on a spot just where the sun would strike the land as it rose above the earth's disk, and sure enough he caught the sun, so that it was held fast in the cord, and did not rise. The animals who ruled the earth were immediately put into a great commotion. They had no light. They called a council to debate upon the matter, and to appoint some one to go and cut the cord, for this was a very hazardous enterprise, as the rays of the sun would burn whoever came so near. At last the dormouse undertook it, for at this time the dormouse was the largest animal in the world. When it stood up, it looked like a mountain. When it got to the place where the sun was snared, its back began to smoke, and burn with the intensity of the heat, and the top of its carcass was reduced to enormous heaps of ashes. It succeeded, however, in cutting the cord with its teeth, and freeing the sun, but it was reduced to a very small size, and has remained so ever since."

In this myth we have the Sun-catcher myth of the South Sea Islands combined with part of our own fairy tale of Jack-and-the-Bean-Stalk. In this latter tale it is related that Jack, in spite of his mother's prayers, goes up the ladder that is to take him to the dwelling of the Giant who killed his father; so the boy of the Indian legend will not heed his sister's persuasion, but goes to seek the enemies who had slain his kindred.

In these myths the loosing of the imprisoned Sun is told in a story of which the European fable of the "Lion and the Mouse" might be a mere moralised remnant.

We have another version of the foregoing myth, which was told by the Wyandot Indians to the missionary Paul Le Jeune:

"There was a child whose father had been killed and eaten by a bear, and his mother by the Great Hare. A woman came and found the child and adopted him as her little brother, calling him 'Chakabech.' He did not grow bigger than a baby, but he was so strong that the trees served as arrows for his bow. When he had killed the destroyers of his parents, he wished to go up to heaven, and climbed up a tree. Then he blew upon it and it grew up and up till he came to heaven and there he found a beautiful country. So he went down to fetch his sister, building huts as he went down to lodge her in, brought her up the tree into heaven, and then broke off the tree low down, so that no one can go up to heaven that way. Then Chakabech went out, and set his snares for game, but when he got up at night to look at them, he found everything on fire, and went back to his sister to tell her.   Then she told him he must have caught the sun.   Going along by night he must have got in unawares, and when Chakabech went to see, so it was, but he dared not go near enough to let the sun out.   By chance he found a little mouse, and blew upon her till she grew so big that she could set the sun free, and the sun released from the trap went again on his way, but while he was held in the snare, day failed down here on earth."

Still another version of the Sun-catcher myth is found among the Dogrib Indians, who dwell in the far North-west:

"When Chapewee after the Deluge formed the earth and landed the animals upon it from his canoe, he stuck up a piece of wood which became a fir-tree, and grew with amazing rapidity until its top reached the skies.   A squirrel ran up this tree, and was pursued by Chapewee, who endeavoured to knock it down, but could not overtake it.   He continued the chase however, until he reached the stars, where he found a fine plain, and a beaten road.   In this road he set a snare made of his sister's hair, and then returned to earth. The sun appeared as usual in the heavens in the morning, but at noon it was caught by the snare

which Chapewee had set for the squirrel, and the
sky was instantly darkened. Chapewee's family,
on this said to him: 'You must have done
something wrong when you were aloft, for we no
longer enjoy the light of day.' 'I have,' replied
he, 'but it was unintentional.' Chapewee then
endeavoured to repair the fault he had committed,
and sent a number of animals up the tree to release
the sun by cutting the snare, but the intense heat
of that luminary reduced them all to ashes. The
efforts of the more active animals being thus
frustrated, a ground mole, though such a grovelling
and awkward beast, succeeded by burrowing under
the road in the sky until it reached and cut asunder
the snare which bound the sun. It lost its eyes,
however, the instant it thrust its head into the
light, and its nose and teeth have ever since been
brown as if burnt."

In the following Omaha myth of "How the
Rabbit Caught the Sun in a Trap,"[1] we find the
Sun ensnared again unwittingly. These myths
differ from the Polynesian Sun-catcher myths in
this respect,—that there appears to have been no
deliberate intention of catching the Sun, no deliberate plan to restrain his liberty, which is a characteristic feature of the South Sea Island myths.

"Once upon a time a rabbit dwelt in a lodge

[1] Annual Report of the Bureau of Ethnology, 1879–80.

with no one but his grandmother. He was accustomed to go hunting early in the morning. Inevitably a person with very long feet had preceded him, leaving a trail. The rabbit desired to find out who this party was, and got up one morning very early, but even then he had been preceded, so he laid a snare that night so as to catch this early bird, laying a noose where the footprints used to be seen. Rising early the next morning he inspected his trap, and found he had caught the sun. Now he was very much frightened, and the sun said to him: 'Why have you done this? You have done a great wrong. Come hither and untie me.' Finally the rabbit mustered up courage and bending his head down rushed at the sun and severed the rope with his knife, but the sun was so hot that the rabbit scorched the hair between his shoulders, so that it was yellow, and from that time the rabbit has had a singed spot on his back between his shoulders."

In the legends of the Bungee Indians of Lake Winnipeg, we find again a reference to the state of dissatisfaction existing in early times with the Sun's vagarious method of lighting the world, and the schemes that were suggested to bring about a change of conditions. In the following myth[1] the Sun is ensnared as the result of a deliberate plan:

[1] Vol. xix., *Journal American Folk-Lore.*

"Before the Creation, the world was a wide waste of water, and there was no light upon the earth, the sun being only an occasional visitor to this world. Anxious to keep the sun from wandering away very far, the god Weese-ke-jak constructed an enormous trap to catch the sun, and the next time the sun came near the earth he was caught in the trap. In vain he struggled to get free, but the cords by which he was held were too strong for him. The near proximity of the sun to the earth caused such heat, that everything was in danger of being burned. Then Weese-ke-jak concluded to make some sort of a compromise with the sun before he would consent to give him his liberty. It was stipulated that the sun was only to come near the outer edges of the earth in the mornings and evenings, and during the day to keep farther away, just near enough to warm the earth without scorching it. But now another difficulty presented itself, the sun had not the power to unloose the band by which he was held, and the intense heat prevented either Weese-ke-jak or any of his creations from approaching the sun to cut the band and set him free. The beaver at that time was rather an insignificant creature, having only a few small teeth in his head, and being covered with bristly hair like a hog, his tail being only a small stump about two or three inches long. He offered to release the sun, and succeeded in gnawing through the cords that held the sun before being quite roasted alive. The cords being severed the sun rose from the earth like a vast balloon. Weese-ke-jak in gratitude for his deliverance from the burning rays of the sun rewarded the beaver by giving him a beautiful soft coat, and fine sharp teeth of a brown colour, as if scorched by fire. This is how the beaver came by his hatchet-like teeth and furry coat."

A feature of these legends is the stress laid on man's indebtedness to a small and insignificant animal, which, in every case, at the risk of his own life, frees the Sun from the toils into which he has been brought by man's machinations. A moral seems to be drawn from these myths, that even the lowly may effect great things, and the despised of earth may rise to heights which even the mighty cannot attain.

We come now to a brief discussion of the myths relating to the temporary retarding or accelerating of the Sun's speed in order that man might accomplish a purpose. Chief among these traditions is the Biblical story of Joshua's command to the Sun to stay its course, related in the tenth chapter of the Book of Joshua: "And he [Joshua] said in the sight of Israel, Sun, stand thou still upon Gibeon; and thou, Moon, in the valley of Ajalon. And the sun stood still, and the moon stayed, until the peo-

Joshua Commanding the Sun

Mosaic, S. Maria Maggiore, Rome

The Bull Apis
Louvre

ple had avenged themselves upon their enemies. Is not this written in the book of Jasher? So the sun stood still in the midst of heaven, and hasted not to go down about a whole day. And there was no day like that before it or after it, that the Lord hearkened unto the voice of a man, for the Lord fought for Israel."

This idea that man could stay the Sun's daily course is to be met with in the Fiji Islands, where, on the top of a small hill, a patch of reeds grew. Travellers who feared that they would not reach the end of their journey before the Sun set, were wont to tie the tops of a handful of these reeds together to detain the Sun from going down. It has been thought that by this act they may have imagined they could entangle the Sun in the reeds for a time and thus stay his course.

When the Australian blackfellow desired to prevent the Sun from going down till he reached home, he placed a sod in the fork of a tree exactly facing the setting Sun. Another Australian custom is to place stones in trees at different heights from the ground, in order to indicate the height of the Sun in the sky at the moment when they passed a particular tree. Those following are thus made aware when their friends in advance passed the spot. Frazer,[1] referring to this custom, considers

[1] *The Golden Bough*, J. G. Frazer.

that the natives who practised it may have imagined that "to mark the sun's progress was to arrest it at the point marked. On the other hand, to make the sun go faster, the Australians throw sand into the air, and blow with their mouths toward the sun."

South African natives in travelling will put a stone in a branch of a tree, or place some grass on the path with a stone over it, believing that this will cause their friends to keep the meal waiting till their arrival, as it would make the Sun go slower down the western sky.

The Indians of Yucatan, when journeying westward, placed a stone in a tree, or pulled out some of their eyelashes, and blew them toward the Sun to stay or speed the Sun's course.

The idea that the Sun's speed could be regulated may have arisen from the fact, that, under certain conditions, it really does appear to vary in its rate of motion in relation to its position in the heavens. During the first few hours succeeding sunrise, when the Sun is not far from the horizon, and can be compared with terrestrial objects, it really appears to move with greater speed than when it journeys across the meridian; and when at nightfall it seeks the west it seems in like manner to hasten with accelerated speed. From this optical illusion the fancy may have sprung born of the false reasoning that, if the Sun varied his speed to suit his will, man could also control his course, and hasten or retard his progress. This has doubtless given rise to the similarity of these legends that have come down to us from many different and widely separated lands, for primitive man the world over viewed the phenomena of nature from much the same standpoint, and wove his legends from the fabric of an imagination common to all men.

Chapter IX
Solar Festivals

Chapter IX

Solar Festivals

BECAUSE of the importance of Sun worship, and its widespread influence upon the primitive inhabitants of the world, there were instituted in honour of the solar deity, in many lands, great festivals and elaborate ceremonials, the traces of which have come down to us in modified form even to this day. Our most important ecclesiastical feast days in fact are but survivals of ancient solar festivals.

Twice in the year the sun apparently changes its course. In midwinter, having reached the lowest point in its path, it turns about and begins to mount the skies; in midsummer, conversely, having attained the highest point it reaches, the sun seems to turn about once more, and descend the steeps of the firmament. These two epochs, the winter and summer solstices as they are called, that mark the sun's annual course, were regarded as supremely important by the ancients and gave rise to great national festivals that were celebrated with pomp and ceremony throughout the ancient world.

At the feast of the winter solstice men testified their gladness at witnessing the return of the all-powerful sun. To the inhabitants of Greenland it meant the early return of the hunting season, and all nations regarded it as a sign that spring-time and harvests were on the way, and the dormant life of the winter season was on the wane.

In many countries this festival season was known as "Yole," or "Yuul," from the word Hiaul, or Huul, which even to this day signifies "the sun" in some languages. From this we get our word "wheel," and the wheel is one of the ancient symbols of the sun, the spokes representing the sun's rays. As we shall see later this symbol was a prominent feature in one of the great solar festivals.

Procopius describes how the men of Thule climbed the mountain tops at the winter solstice, to catch sight of the nearing sun after their thirty-five days of night. Then they celebrated their holiest feasts.

Plutarch, referring to the solar festivals of Egypt, says, that "about the winter solstice they lead the sacred cow seven times in procession around the temple, calling this the searching after

Osiris, that season of the year standing most in need of the sun's warmth."

In China, the Great Temple of the Sun at Pekin is oriented to the winter solstice, and the most important of all the State observances of China takes place there December 21st, the sacrifice of the winter solstice.

In our own time a number of Christian religious observances and festivals are of distinct solar origin. Notable among these feast days is Christmas. "The Roman winter solstice," says Tylor,[1] "as celebrated on December 25th (VIII Kal. Jan.) in connection with the worship of the Sun-God Mithra appears to have been instituted in this special form by Aurelian about A. D. 273, and to this festival the day owes its apposite name of 'Birthday of the Unconquered Sun.' With full symbolic appropriateness, though not with historical justification, the day was adopted in the western church where it appears to have been generally introduced by the fourth century, and whence in time it passed to the eastern church as the solemn anniversary of the Birth of Christ, Christmas Day. As a matter of history no valid or even consistent early Christian tradition vouches for it."

Many of the early dignitaries of the Church

[1] *Primitive Culture*, Edward B. Tylor.

reveal in their writings the solar character of this festival. Augustus and Gregory discoursed on "the glowing light and dwindling darkness that follow the nativity," and Leo the Great denounced in a sermon the idea that Christmas Day is to be honoured, not for the birth of Christ, but for the rising of the new sun.

The solar origin of the great feast is attested in Europe by bonfires, and the burning of the Yule log, and in the Christmas service chant, "Sol novus oritur."

Even the sacrifices offered to the Sun in pagan times at the great solar festivals find their survival in the sacrifices of a lamb which we offer at Easter-tide, and an ox at Christmas. The lighting of the Christmas tree is but the light to guide the Sun-God back to life, and the festival cakes of corn and fruit, made in honour of the Sun in ancient times, and laid on the sacred altars of the Persians as an offering of gratitude to the Lord of Light and Life, find their prototype in the plum pudding that graces the board at our Christmas feasts of rejoicing. Christmas is, therefore, nothing but an old heathen celebration of the winter solstice, the feast of re-joicing that a turning point in the sun's course has been reached, and that the life-giving orb has attained the end of its journey of dwindling hours of daylight, and has started back on a course that brings with it each day an increase of warmth and light.

Of equal importance to the solar Christmas festival celebrated at the winter solstice was the great celebration of the summer solstice recognised throughout Europe. This was preëminently a fire festival, for the ceremonies featured the lighting of huge bonfires on the hilltops, leaping through the flames, and rolling blazing wheels of fire from the summits of the hills, indicating the sun's descending course in the heavens.

According to Tylor,[1] "These ancient rites attached themselves in Christendom to St. John's eve. It seems as though the same train of symbolism which had adapted the midwinter festival to the nativity, may have suggested the dedication of the midsummer festival to John the Baptist, in clear allusion to his words, 'He must increase, but I must decrease.'"

Durandus, speaking of the rites of the Feast of St. John the Baptist, informs us of the curious custom that prevailed of rolling fire wheels down from the hills. This practice was common in France, and many North German examples of it are on record. The following is an account of one of these festivals, which took place at Conz on the Moselle, in 1823, as described by Grimm:

[1] *Primitive Culture*, Edward B. Tylor.

"Every house delivers a sheaf of straw on the top of the Stromberg, where the men and lads assemble towards evening, whilst the women and girls gather about the Burbacher fountain. A huge wheel is now bound round with straw in such a manner that not a particle of wood remains visible; a stout pole is passed through the middle of the wheel, and the persons who are to guide it lay hold on the ends of the pole, which projects three feet on either side. The rest of the straw is made up into a great number of small torches. At a signal from the mayor of Sierk (who, according to ancient custom, receives a basket of cherries on the occasion), the wheel is kindled with a torch and set rapidly in motion. Everybody cheers and swings torches in the air. Some of the men remain above, others follow the burning wheel down hill in its descent to the Moselle. It is often extinguished before it reaches the river, but if it burns at the moment it touches the water, that is held to be prophetic of a good vintage, and the people of Conz have a right to levy a fuder of white wine upon the surrounding vineyards. Whilst the wheel is passing before the female spectators, they break out into cries of joy, the men on the hilltops reply, and the people from the neighbouring villages who have assembled on the banks of the river mingle their voices in the general jubilee."

There is a striking analogy between the St. John fire celebrations and the Vedic legend of Indra's fight with the midsummer demons.

"In this legend," says Keary,[1] "the demon Vritra possessed himself of the sun wheel and the treasures of heaven, seized the women, kept them prisoners in his cavern, and laid a curse on the waters until Indra released the captives and took off the curse."

The significance of the ceremony lies in the details that enter into it, the key to which is found in the following passage from a Vedic hymn: "With thee conjoined, O Indu (Soma), did Indra straightway pull down with force the wheel of the sun that stood upon the mighty mountain top, and the source of all life was hidden from the great scather."

The German custom is therefore seen to be nothing but a dramatic portrayal of the great elemental battle as depicted in the sacred books of the ancient Hindus. The wheel of fire on the hilltop represents the sun resting on the crest of the cloud mountain. Both the wheel and the sun descend from their positions of prominence and are extinguished, the wheel by the waters of the stream at the base of the hill, the sun by the sea of clouds.

[1] *Curiosities of Folk-Lore*, C. F. Keary.

The elements of strife and warfare enter into the scene. The descending wheel is pursued to the water's edge by a crowd of men brandishing torches; Indra and his hosts wage successful warfare against the army of the demon Vritra. The fact that women are excluded from the ceremonies emphasises the idea of a combat, for it is their province solely to watch the battle as spectators and cheer the victors.

Another notion associated with this rite of the blazing wheel was that, as the wheel went rolling away from them in its descending course, it symbolised a wheel of fortune, and the ill luck of the people went rolling away, a signal for great rejoicing.

This ceremony of the descent of the wheel was anciently observed on St. John the Baptist's Day at Norwich, England, and even to this day it is the custom to light huge bonfires on the hilltops in Ireland, according to the ancient pagan usage when the Baal fires were kindled as part of the ritual of Sun worship. Around these fires the peasants dance, and when the fire burns low, it is the custom to lift children across the glowing embers to secure them good luck during the year, which is similar to the custom practised by worshippers of Baal and Moloch in ancient times of passing children through the fire that burned at the feet of the cruel and insatiable god.

There was also practised in Ireland, in connection with the midsummer festival which celebrated the turning-point of the sun at the summer solstice, a strange dance which was religious in its character, and solar in its origin. The Greeks called this "the Pyrrhic dance" from "pur" meaning fire, and practised it from the most ancient times. The feature of the dance was its serpentine character, as the dancers circled about in a long line simulating the coils of a serpent. In Ireland the dance had the same characteristics, and though the esoteric meaning of the dance had been lost, it was in all probability a mystic rite symbolic of the course of the sun, for the dancers invariably circled from east to west.

In Wales, the custom of lighting bonfires on Midsummer Eve is still kept up in many villages, and the peasants gather about them dancing and leaping through the flames. The leaping through the flames is supposed to ward off evil spirits, prevent sickness, and bring good luck.

The connection of the ceremony of the bonfires with the old worship of the Sun is indisputable. Its practice was general among nearly all European nations, and in not very remote times, from Norway to the shores of the Mediterranean, the glow of St. John's fires might have been seen.

In Brittany, the custom of the Baal fires is still preserved, and the peasants dance around them all night in their holiday attire. It is said that the maid who dances round nine St. John's fires before midnight is sure to be married within the year. In many parishes the curé himself goes in procession with banner and cross to light the sacred fire, and all the ancient superstitions connected with the festival are kept alive with unabated zeal.

The Scandinavians believed that when midsummer came the death of their Sun-God Balder took place, and to light him on his way to the underworld they kindled bright fires of pine branches, and when, six months later at the winter solstice, he regains his life and mounts to greet them, they burn the yule log and hang lights on the fir-trees to illuminate his upward course.

Frazer tells us[1] how the fern seed, the oak, and the mistletoe are closely associated as symbols with the solar festivals celebrated at the winter and summer solstices:

"The two great days for gathering the fabulous fern seed, which is popularly supposed to bloom like gold or fire on Midsummer Eve, are Midsummer Eve and Christmas, that is, the two solstices. We

[1] *The Golden Bough*, J. G. Frazer.

are led to regard the fiery aspect of the fern seed as primary, and its golden aspect as secondary and derivative. Fern seed, in fact, would seem to be an emanation of the sun's fire at the two turning-points of its course, the summer and winter solstice. This view is confirmed by a German story, in which a hunter is said to have procured fern seed by shooting at the sun on Midsummer Day at noon. Three drops of blood fell down, which he caught in a white cloth, and these blood drops were the fern seed. Here the blood is clearly the blood of the sun from which the fern seed is thus directly derived. Thus it may be taken as certain that the fern seed is golden because it is believed to be an emanation of the sun's golden fire.

"Now, like the fern seed, the mistletoe is gathered either at midsummer or Christmas, that is, at the summer and winter solstice, and, like the fern seed, it is supposed to possess the power of revealing treasures in the earth. Now if the mistletoe discovers gold, it may be in its character of the Golden Bough, and if it is gathered at the solstices, must not the Golden Bough, like the golden fern seed, be an emanation of the sun's fire? The primitive Aryans probably kindled the midsummer bonfires as sun charms, that is, with the intention of supplying the sun with fresh fire. But as the fire was always elicited by the friction of oak wood, it must have appeared to the primitive Aryan that the sun was periodically recruited from the fire which resided in the sacred oak. In other words, the oak must have seemed to him the original storehouse or reservoir of the fire which was from time to time drawn out to feed the sun. Thus, instead of saying that the mistletoe was an emanation of the sun's fire, it must be more correct to say that the sun's fire was regarded as an emanation of the mistletoe."

The Christian festival of Easter has its solar characteristics. "The very word Easter," says Proctor, "is in its real origin as closely related to sun movements as the word East," and the notion that the Sun dances on Easter morning as it rises is firmly believed to-day by superstitious people. In Saxony and Brandenburg the peasants still climb the hilltops before dawn on Easter day to witness the three joyful leaps of the Sun, as our English forefathers used to do.

Tylor[1] tells us that "the solar rite of the New Fire, adopted by the Roman Church as a Paschal ceremony, may still be witnessed in Europe with its solemn curfew on Easter Eve, and the ceremonial striking of the new holy fire."

The two great festivals of the ancient Irish were La Baal Tinné, or May Day, the day of the Baal

[1] *Primitive Culture*, Edward B. Tylor.

fires, sacred to the Sun, and La Samnah, or November Eve, sacred to the Moon. The May festival was the most important, as then it was that the Druids lit the fire of Baal, the Sun-God, and a portion of the ceremony at this festival consisted in driving cattle along a narrow path flanked by two fires, singeing them with the flame of a torch, and sometimes bleeding them, the blood being then offered as a sacrifice to the Sun-God.

Plutarch relates that among the Egyptians there were several festivals in honour of the Sun. A solar sacrifice was performed on the fourth day of every month, and so important was the deity that, as propitiatory offerings, they burnt incense three times a day, resin at its first rising, myrrh when on the meridian, and a mixture called "kuphi" at sunset. A festival in honour of the Sun was held on the thirtieth day of Epiphi, called the "birthday of Horus' eyes," when the sun and moon were supposed to be in the same right line with the earth. On the twenty-second of Phaophi, after the autumnal equinox, there was a similar ceremonial to which, according to Plutarch, they gave the name of the "nativity of the staves of the sun," intimating that the sun was then removing from the earth, and as its light became weaker and weaker, that it stood in need of a staff to support it.

The most important date of the Egyptian year was the twentieth of June, that marked the summer solstice, but more especially the rise of the all-fertilising Nile. This was the New Year Day in Egypt. The greatest solar festival of the Egyptians, however, was the festival of Osiris, and the special feature of this occasion was the procession in which the sacred ox Apis appeared.

The following description of this festival is taken from *Mythology and Fables* by the Abbé Banier:

"The ox whom the priests nourished with so much care, and for whom all Egypt had such a veneration, was looked upon as a god. To gain some credit to this superstition, they said he represented the soul of Osiris. Herodotus tells us that this ox was to be black over all the body, with a square white mark upon the forehead. Upon the back he was to have the figure of an eagle, a knot under the tongue in the figure of a beetle, the hairs of the tail double, and according to Pliny a white mark upon the right side, which was to resemble the crescent moon. Porphyry says that all these marks had reference to the sun and moon, to whom the ox Apis was consecrated, that the black hair which was to be the colour of his body in general represented the scorching influence of the sun upon bodies, and that the white spot which he had in his forehead, and the crescent which he bore upon the side, were symbols of the moon. The eagle and beetle were also symbols of the sun.

"The festival of Apis lasted seven days. The people went in crowds to bring him from the place where he was found, the priests led the procession, and every one was desirous to receive him into his home. On the day of the Osiris festival the priests conducted the ox Apis to the banks of the Nile and drowned him with great ceremony. He was then embalmed and interred at Memphis. After his death the people mourned and made lamentation as if Osiris had been now dead. The priests cut off their hair, which in Egypt was a sign of the deepest mourning, and this mourning lasted till they got another ox to appear resembling the former in the same marks, when they began to make merry as if the Prince himself had arisen from the dead. The superstition of the Egyptians in relation to the ox Apis was carried to great excess. They honoured him as a god, and consulted him as an oracle; when he took what food was offered to him it was a favourable response, and his refusing it was looked upon as a bad presage."

In Greece there were many solar festivals inaugurated in honour of the Sun-God Phœbus Apollo. In Sparta an annual festival, known as "Carneia," was held in August. It was a religious ceremony, the purpose of which was to appease the dreaded god. This festival was celebrated in Cyrene, in the islands of Rhodes and Sicily, and in many of the Greek cities in lower Italy.

In September a festival was celebrated at which the Sun-God was invoked as an aid in battle, and in October the first-fruits were presented as a sacrifice.

At Athens there was an annual festival, held in May, to commemorate the yearly tribute of youths and maidens to Crete as sacrifices to the Minotaur. At Thebes there was a festival in honour of Apollo Ismenius, held every eighth year, called the "Daphnephoria." At this celebration branches of olives hung with wreaths, and representations of the sun, moon, and stars, were carried in procession, a feature of the festival.

A festival in honour of "Hyacinthus," one of the titles of Apollo, was celebrated annually at Sparta, in July, and lasted nine days. It began with laments, but concluded with expressions of joy and gladness. In honour of Apollo, the Sun-God, a festival called "Thargelia" was held at Athens, in May, to celebrate the harvest yield. In August the Athenians celebrated a similar festival called "Metageitnia." These celebrations and festivals bear testimony to the importance of Sun worship among the ancient Greeks.

Among the Peruvians there were four solar festivals of importance celebrated throughout the year with great pomp and display. Chief among these was that of the winter solstice, which fell in June. It was the festival of the diminished and growing Sun. It lasted nine days, the first three of which were given up to fasting. On the morning of the great day the Emperor himself officiated as high priest, and all the people gathered at dawn in the public square to await the coming of the supreme deity, the Sun. At sight of him great shouts of joy rose from the multitude, who threw kisses to the orb of day, and prostrated themselves. The chief priest then offered a libation to the Sun-God, drinking of the cup himself, and then passing it on to his retinue, an act of solar communion. All then marched to the temple of the Sun, where a black llama was sacrificed, and its entrails were carefully inspected for omens affecting the coming year. A fire, produced by the focused rays of the sun from a mirror, was then lighted on the altar, and from it fire was conveyed to all the Sun temples in the city. These fires were kept burning continuously until three days before the next solstice when they were allowed to burn out.

The second great solar festival of the Peruvians was known as the "feast of purification," and fell in September. The object of this ceremonial was to invoke the Sun's aid and beneficent influence to secure the prosperity, health, and security of the people. The third festival was held in May, and was a thanksgiving harvest celebration, while still a fourth festival took place in December known as the "festival of power."

The Hopi Indians of North America, in their elaborate festivals in honour of the Sun, impersonated the Sun-God. The impersonator wore a disk-shaped mask, surrounded with eagle-wing feathers, and this was fringed with flowing strands of red horsehair to represent the sun's rays. The sun masks were a prominent feature in the solar ceremonials of many of the Indian tribes.

The curious and interesting custom of "need-fires," although not exactly to be classed as solar festivals, may very properly be treated of in this chapter, owing to the solemnity of the ceremony, the implicit faith of the people in their efficacy, and especially owing to their solar significance.

The "need-fires," or "forced-fires" as they are sometimes called, were kindled at times of great epidemics among the cattle that threatened their total annihilation, and the custom of kindling these fires is still in vogue in certain countries. Keary[1] thus describes the custom:

"Wherever it can be traced among people of

[1] *Curiosities of Folk-Lore*, C. F. Keary.

Title: Young titan : the making of Winston
Churchill
ID: 33100005795827
Due: 6/13/2013,23:59

Just checked out: 1
5/16/2013 11:32 AM
Total Checked out: 4
Overdue: 0
Hold requests: 0
Holds ready for pickup: 0

German or Scandinavian descent, the fire is always kindled by the friction of a wooden axle in the nave of a waggon wheel or in holes bored in one or two posts. In either case the axle or roller is worked with a rope, which is wound around it and pulled to and fro with the greatest possible speed by two opposite groups of able-bodied men. The wheel was beyond all doubt an emblem of the sun. In a few instances of late date it is stated that an old waggon wheel was used.

"In Marburger, official documents of the year 1605, express mention is made of new wheels, new axles, and new ropes, and these we may be assured were universally deemed requisite in earlier times. It was also necessary to the success of the operation that all the fires should be extinguished in the adjacent houses, and not a spark remain in any one of them when the work began. The wood used was generally that of the oak, a tree sacred to the lightning god Thor, because of the red colour of its fresh-cut bark. Sometimes, especially in Sweden, nine kinds of wood were used   The fuel for the fire was straw, heath, and brushwood, of which each household contributed its portion, and it was laid down over some length of the narrow lane which was usually chosen as the most convenient place for the work. When the fire had burned down sufficiently, the cattle were forcibly driven through it two or three times in a certain order beginning with the swine, and ending with the horses or vice versa. When all the cattle have passed through the fire, each householder takes home an extinguished brand which in some places is laid in the manger. The ashes were scattered to the winds apparently that their wholesome influence may be spread far abroad. In Sweden the smoke of the need-fires was believed to have much virtue: it made fruit trees productive, and nets that had been hung in it were sure to catch much fish.

"The earliest account of the need-fires in England is that quoted by Kemble from the Chronicle of Lanercost for the year 1268. The writer relates with pious horror how 'certain bestial persons, monks in garb but not in mind, taught the country people to extract fire from wood by friction, and set up a "Simulacrum Priape" as a means of preserving their cattle from an epidemic pneumonia.' This 'Simulacrum Priape' was unquestionably an image of the sun-god Fro or Fricco.

"Jacob Grimm was the first to make it evident that, for the Germans at least, the wheel was an emblem of the sun, and numerous facts which have come to light since he wrote, abundantly verify his conclusion.

"He mentions among other evidence that, in the Edda, the sun is called 'fagrahvel,' 'fair, or bright wheel,' and that the same sign ☉ which in the calendar represents the sun stands also for the Gothic double consonant 'hw,' the initial of the Gothic word 'hvil,' Anglo-Saxon 'hveol,' English 'wheel.'

"In the need-fires on the island of Mull, the wheel was turned, according to Celtic usage, from east to west, like the sun. Grimm has also noticed the use of the wheel in other German usages as well as in the need-fire, and he is of opinion that in heathen times it constantly formed the nucleus and centre of the sacred and purifying sacrificial flame. There was a twofold reason for this use of the emblem of the sun, for that body was regarded not only as a mass of heavenly fire, but also as the immediate source of the lightning. When black clouds concealed the sun, the early Aryans believed that its light was actually extinguished, and needed to be rekindled. Then the pramantha[1] was worked by some god in the cold wheel until it glowed again, but before this was finally accomplished the pramantha often shot out as a thunderbolt from the wheel, or was carried off by some fire robbers. The word 'thunderbolt' itself like its German equivalents expresses the cylindrical or conical form

---

[1] The pramantha was the handle of the mythical hand-mill of Frode, the regent of the Golden Age. This hand-mill was a flat circular stone which represented the sun's disk, its handle was used by Indra and the Aswins to kindle the cloud-extinguished orb of day.

of the pramantha.　When the bolts had ceased to fly from the nave, and the wheel was once more ablaze, the storm was over."

From the foregoing, which treats merely of the more important solar festivals, it is clear that these products of paganism are as much in force at present from a symbolic point of view, as they ever were, and that Christianity countenances, and in many cases has actually adopted and practises, pagan rites whose heathen significance is merely lost sight of because attention is not called to the sources whence these rites have sprung.

In short, Sun worship, symbolically speaking, lies at the very heart of the great festivals which the Christian Church celebrates to-day, and these relics of heathen religion have, through the medium of their sacred rites, curiously enough blended with practices and beliefs utterly antagonistic to the spirit that prompted them.

The reason for the survival of many of the symbols of Sun worship and the practice of many customs peculiar to this ancient form of idolatry, lies in the fact that the early Christian teachers found the people so wedded to their old rites and usages, that it was vain to hope for the complete abandonment of these long-cherished practices. Hence a compromise was wisely effected, and the old pagan customs were deprived of the idolatry that was so obnoxious to the Christian, and transferred as mere meaningless symbols and empty forms to the Christian festivals. Old paganism died hard, and fought long and stubbornly in its struggle with Christianity, but time has fought for the Christian, and now even the meaning of symbols and forms that once played such an important part in pagan worship is lost sight of, and their former force and power is lost for evermore.

Chapter X
Solar Omens, Traditions, and Superstitions

Chapter X

Solar Omens, Traditions, and Superstitions

THE past has bequeathed to us a wealth of lore, in the nature of omens and wise saws, superstitions and quaint fancies, the product of imaginative speculation in every phase of human existence, from the earliest times. This is especially true of the influence exerted on human affairs by the sun, moon, and stars; and though, as might be expected, less superstition is attached to the sun than to the moon, owing to the fact that the latter rules the night, when the imagination is roused to activity by the deep shadows and the mysterious gloom, still there cluster about the sun many curious ideas, that, apart from their value to the antiquarian, are interesting to the layman, affording as they do an insight into the life of ancient times.

These superstitions naturally relate to the widest possible range of subjects, for, once given the idea, prevailing at one time in the world's history, that the sun was a living being, who, if he did not rule man's destiny, still had a great share in shaping or controlling it, there is no end to the play of the imagination respecting its influence on man's daily existence. Therefore, what follows must necessarily be fragmentary in its nature, impossible of extensive classification, and consequently disconnected. The purpose is merely to place before the reader the better known and most popular of the fancies relating to the sun that primitive and simple-minded people of past ages created, and which, because of the hold which superstition has ever exerted on the race, are believed in, even in this enlightened age.

In the Ægean Islands, a land teeming with myth and legend, there are still extant many strange superstitions respecting the sun. When the Sun disappears from sight in the west each night, they say he has returned to his vast kingdom in the underworld, where he dwells in a great castle. His mother waits to receive him, and has forty loaves of bread ready to appease his appetite; but if, by any chance, this meal is not prepared, the famished Sun becomes a cannibal, and eats his entire family. When he rises red, the islanders say: "He has eaten his mother. He is crimson-hued because he yielded to his bloodthirsty inclinations, when he found no bread to eat."

Others say the red colours at sunset are caused by the blood flowing from the Sun-God when he hastens to his suicide. Curiously enough, the Greeks regard the rising sun as ushered in through the portals of the east each morning by the Virgin Mary. Again, the sun was regarded by the Greeks as the symbol of perfect beauty, and they formerly painted the sun's disk on the cheeks of a bride. Churches are still dedicated to the Virgin beautiful as the sun, and there are many legends which relate to maidens who boasted that their beauty excelled that of the sun, and of the penalty they paid for making such presumptuous claims.

The Greeks regarded the orb of day as an all-seeing eye, and believed that no deed escaped its detection. The Finns believed that even the abode of the dead could be reached by the blissful rays of the sun.

Because the sun looked down on all men, messages were given to the Sun that he might convey them to absent ones of a family, whom he beheld wherever they chanced to be.

We have seen how the Greeks explained the crimson hues that accompanied the rising sun. Its ruddy hue at sunset also called for an explanation, and the ancients believed that, as the sun reached its vanishing point, it gazed on the fires of hell, and these lit up its face, and the western sky.

It seems to have been an almost universal belief among primitive people that the sun and moon were the abodes of departed souls. In Isaac Taylor's *Physical Theory of Another Life*, we read that "the sun of each planetary system is the house of the higher and ultimate spiritual corporeity, and the centre of assembly to those who have passed on the planets their preliminary era of corruptible organisation."

One of the most popular solar superstitions, and one that has survived even to this day, is the notion that the sun dances when it rises on Easter Day. In the middle districts of Ireland the peasants rise at an early hour Easter morning to witness this phenomenon, which they say is in honour of the resurrection. Brand[1] tells us this "is not confined to the humble labourer and his family, but is scrupulously observed by many highly respectable and wealthy families."

Sir Thomas Browne has left us the following quaint thoughts on this subject: "We shall not, I hope," says he, "disparage the Resurrection of our Redeemer, if we say that the sun doth not dance on Easter Day, and though we would willingly assent unto any sympathetical exultation, yet we cannot conceive therein any more than a topical expression. Whether any such motion there was

[1] *Popular Antiquities*, John Brand.

in that day Christ arised, Scripture hath not revealed, which hath been punctual in other records
concerning solitary miracles, and the Areopagite
that was amazed at the eclipse, took no notice of
this; and if metaphorical expressions go so far,
we may be bold to affirm, not only that one sun
danced, but two arose on that day; that light
appeared at his nativity, and darkness at his death,
and yet a light at both; for even that darkness was
a light unto the Gentiles, illuminated by that
obscurity.   That was the first time the sun set
above the horizon.   That, although there were
darkness above the earth, yet there was light
beneath it, nor dare we say that Hell was dark if
he were in it.''

In some parts of England this quaint belief in
the dance of the joyful Easter sun was regarded
as the lamb playing for very gladness, in honour of
the risen Christ.

In a rare book entitled *Recreation for Ingenious
Head Pieces*, the Easter sun dance is thus referred
to in an old ballad:

> "But, Dick, she dances such a way
> No sun upon an Easter day
> Is half so fine a sight."

The Swabian people are firm believers in this
superstition, and aver that the Sun leaps thrice for

joy when it rises Easter morning; but at Rotten-burg on the Neckar, the Sun was supposed to indulge in this dance at his setting Christmas Eve.

In England, we find this belief in the Easter sun dance still extant in parts of Yorkshire, Durham, Northumberland, and Devonshire, and it is the custom for the maidens to rise early Easter morning that they may witness the sight, and they also look for the lamb and flag in the centre of the sun's disk.

There is little doubt that the northern European nations welcomed the return of the spring sun with dancing, and the May rejoicings familiar in Cornwall are but an expression of congratulation to the spring. We have a similar custom in this country to-day, and May Day is a festival cele-brated with elaborate Maypole dances by the school children of New York City. We see a striking analogy in these dances to the sun dances of the American Indians, which were part of the ritual of their Sun worship, and the radiating lines of ribbons from the Maypole represent the rays of the sun, as the thongs attached to the sun-pole in the Indians' dance did.

The old Beltane games and dances, common in Perthshire and other parts of Scotland until the beginning of the last century, had a solar signifi-cance, the word "Beltane" being a derivative of

Photo by Mooney

Medicine Lodge. Sioux Indian Sun Dance

Courtesy of Smithsonian Institution, Bureau of American Ethnology

The Maypole Dance

the compound word "Baal," the Phœnician word for sun, and "tein," the Gælic word for fire.

Lady Wilde[1] says that "the Beltane dance, where the participants circle about a bush hung with ribbons and garlands, or about a lighted bush or bonfire, celebrating the returning power of the sun, is still kept up in parts of Ireland on May Day and that those taking part always move sunwise."

This sunwise motion is found in many customs extant to-day. In the Scotch Highlands they still "make the deazil" around those whom they wish well of. This superstition consists in walking three times around the person according to the course of the sun. To circle in the opposite direction or "withershins," is productive of evil, and brings bad luck. We see a survival of this custom of circling about to bring luck in the modern superstition often practised by a card-player to-day, when he rises and walks around his chair three times to produce good luck.

According to an Icelandic saga, a woman going against the sun round a house, and waving a cloth, brought down a landslip against the house, and in Yorkshire it is said that if you walk round the room against the sun at midnight, in perfect darkness, and then look into a mirror, you will see leering out of it at you the face of the devil.

[1] *Ancient Legends of Ireland*, Lady Wilde.

In *Popular Romances of the West of England*, by Robert Hunt, there is the following description of a superstitious belief that the sun never shines on a perjured person: "When a man has deeply perjured himself, especially if by his perjury he has sacrificed the life of a friend, he not merely loses the enjoyment of the sunshine, but he actually loses all consciousness of its light or its warmth. Howsoever bright the sun may shine, the weather appears to him gloomy, dark, and cold. I have recently been told of a man living in the western part of Cornwall, who is said to have sworn away the life of an innocent person. The face of this false witness is said to be the colour of one long in the tomb, and he has never, since the death of the victim of his forswearing, seen the sun. It must be remembered the perjured man is not blind, all things round him are seen as by other men, but the sense of vision is so dulled that the world is forever to him in a dark vapoury cloud."

Among the Tunguses, an accused man has to walk toward the sun brandishing a knife, and crying: "If I am guilty, may the sun send sickness to rage in my bowels like this knife."

The appearance of three suns, it is said, denotes war. It is claimed that they are only visible at sunrise, and differ in size. At Herbertingen, they aver that these three suns have frequently been seen, and that they appeared just before Napoleon's disastrous campaign in Russia. The largest sun in this case was in the northern direction, and they say that is why the Russians triumphed.

The sun is also an important factor in Mexican superstition. The following cases are typical, and their injunctions closely followed by the people of that country:

The head of the bed must never be placed toward the rising sun, since it will cause the sleeper to rise with a bad headache, and even insanity may result. The sun is also appealed to whenever a tooth drops out, or is extracted. When this occurs, the loser takes the tooth and throws it with all his might at the sun.

When the sun sets on a cloudy day, the following day will be a stormy one. The Mexicans also have a belief that blondes cannot see the sun.

In Germany it is the custom on St. John's Day for hunters to fire at the sun, believing that they will thereby become infallible hunters. According to another German belief, he who on St. John's Day fires toward the sun is condemned to hunt forever afterward, like Odin, the eternal hunter.

We are all familiar with the phrase, "the sun is drawing water," used to express the appearance of the sun's rays as they filter through spaces in the clouds, and spread out like a fan over a body of water. This expression arose from the fact that primitive people fancied the sun's rays on these occasions resembled ropes that ran over the pulleys of the old-fashioned draw wells.

A very strange belief is that of the Namaqua Hottentots, that the sun is a bright piece of bacon, which the people who go in ships draw up in the evening by enchantment, and let down again after they have cut a piece off from it.

There is a curious custom found in many parts of the world, which relates to the sun's influence on young maidens entering on womanhood. According to this superstition, these maidens must not touch the ground nor permit the sun to shine upon them. In Fiji, brides who were being tattooed were hidden from the rays of the sun, and in a modern Greek folk-tale the Fates predict that in her fifteenth year a princess must be careful not to let the sun shine on her lest she be turned into a lizard. A Tyrolese story tells how it was the doom of a lovely maiden to be transported into the belly of a whale if ever a sunbeam fell on her.

The old Greek legend of Danaë, who was imprisoned by her father in a dungeon, or brazen tower, is a further illustration of this strange idea relating to the sun's influence on human affairs.

The following solar superstitions show the wide

play of man's fancy as relating to the power of the sun's light and indicate that there was a belief that the sun was a devout Christian, and kept the holy days prescribed by the Church:

For the sun to shine upon a bride is a good omen.

> "While that others do divine
> Blest is the bride on whom the sun doth shine."
> HERRICK'S *Hesperides*.

If the sun shines while it rains, the witches are baking cakes.

The Mexicans say when it rains, and the sun is shining, a she-wolf is bringing forth her offspring, or a liar is paying his debts.

If the sun shines on Candlemas Day (February 2d), the flax will prosper.

If women dance in the sun at Candlemas their flax will thrive that year.

> "As far as the sun shines in on Candlemas Day
> So far will the snow blow in before May:
> As far as the snow blows in on Candlemas Day
> So far will the sun shine in before May."

> "If the sun in red should set,
> The next day surely will be wet;
> If the sun should set in grey,
> The next will be a rainy day."

When the sun does not shine, all treasures buried in the earth are open.

If the sun shines on Easter Day, it will shine on Whitsunday also.

On Good Friday the sun mourns over the Crucifixion, and does not shine until three o'clock in the afternoon.

The sun is obliged to shine for a short time at least every Sunday in order that the Blessed Virgin may dry her veil.

Three Saturdays in the year, when the Virgin Mary mourns, the sun does not shine at all.

From an old Dream Book we derive the following superstitions:

To dream you see the sun shine, shows accumulation of riches and enjoying posts of honour in the state, also success to the lover.

To dream you see the sun rise, promises fidelity in your sweetheart, and good news from friends.

To dream you see the sun set, shows infidelity in your sweetheart, and disagreeable news. To tradesmen, loss of business.

To dream you see the sun under a cloud, foretells many hardships and troubles are about to befall you, and that you will encounter some great danger.

Chapter XI
Solar Significance of Burial Customs. Orientation

Chapter XI

Solar Significance of Burial Customs. Orientation

NOWHERE in the study of ancient rites and customs is the sun's influence on human affairs in greater evidence than in the ceremonials attending the burial of the dead.

The funeral rites of all people reveal the universal belief that the east is the source of all that men hold dear, light, life, warmth, and happiness, while the west, on the contrary, is said to be the abode of darkness, death, cold, and sorrow. The worship of the Sun cultivated and strengthened this idea, and down through the ages the influence of this belief has swept, retaining even to-day much of its ancient force and vigour.

According to Tylor[1]: "It seems to be the working out of the solar analogy on the one hand in death at sunset, on the other in new life at sunrise, that has produced two contrasted rules of burial

---

[1] *Primitive Culture*, Edward B. Tylor.

which agree in placing the dead in the sun's path, the line of east and west."

It is said that the body of Christ was laid with the head toward the west, that the risen Lord might face the eastern realm of eternal life and glory, and the Christian custom that sprang from this belief led to the usage of digging graves east and west, which prevailed through mediæval times, and is common with us to-day.

In the twenty-fourth chapter of St. Matthew's gospel we read: "For as the lightning cometh out of the east, and shineth even unto the west; so shall also the coming of the Son of man be." From the literal interpretation of these words there arose the belief that Jesus would, at the resurrection, appear from the east, and hence that those buried with their faces upward and their heads to the west, would be in readiness to stand up with their faces toward their Judge.

Swift alludes to this custom in the account of Gulliver's voyage to Lilliput, where he says: "The inhabitants bury their dead with their heads directly downward because they hold an opinion, that in eleven thousand moons they are all to rise again, in which period the earth, which they conceive to be flat, will turn upside down, and by this means they shall at their resurrection be found ready standing on their feet."

Sir Walter Raleigh referred to this superstition when he stood on the scaffold, and was about to be executed. After forgiving his executioner, there was a discussion as to the way he should face, some saying he should face the east. Raleigh then remarked: "So the heart be straight it is no matter which way the head lieth."

The east and west burial custom was practised by the ancient Greeks, and by the natives in some districts of Australia, although the latter people as a rule regarded the west as the abode of departed spirits, and therefore buried their dead facing that quarter.

The native Samoans and Fijians follow the same custom, believing that if the dead are buried with head east and feet west, the body at the resurrection would be in a position to walk straight onward to the abiding-place of the soul.

According to Schoolcraft, the Winnebago Indians buried their dead in a sitting posture with the face west, or at full length with the feet west, in order that they may look toward the happy land in the west. Other Indian tribes, notably the Indians of Kansas, practised this custom.

It was the Peruvian custom to bury the dead huddled up in a sitting posture with their faces turned toward the west, and in the funeral ritual of the Aztecs there is found a description of the first peril that the shade encountered on its journey to the abode of the dead, which they believed was illuminated by the sun when night enveloped the earth.

On the contrary, the Yumanas of South America were accustomed to bury their dead in a sitting posture facing the east, as they believed that in the east was the home of their supreme deity, who would one day take unto himself all true believers in him. The Guayanos have a similar belief and custom. The modern Ainus of Yezo bury their dead lying robed in white with heads to the east, because that is where the sun rises. The mediæval Tartars raised a great mound over their graves, and placed therein a statue with its face turned eastward.

The Siamese believe that no one should sleep with his head to the west, as that is the position in which the dead are placed at burial. Lady Wilde calls attention to the curious customs, practised at the wake of the Irish peasants, which are derived from the ancient funeral ceremonies of the Egyptians. Particularly is this the case where, during the wake, a man and a woman appeared, one bearing the head of an ox, the other that of a cow. This strange custom is thought to represent the Egyptian gods Isis and Osiris waiting to receive the soul of the dead.

In Wales the east wind is called the "wind of the dead men's feet," and the eastern portion of a churchyard was always regarded as the most honoured part. South, west, and north were next in favour, in their order, and suicides were buried with their heads to the north, as, in taking their own lives, they had forfeited the rights of the orthodox to a burial with face to the east. In rural parts of England it was the custom in ancient times to remark at the funeral service: "The dead ay go wi' the sun."

Even in our own country we see a survival of the universal belief in the proper orientation of a deceased person. Examination shows that the headstones in the old burial-grounds of Plymouth, Concord, and Deerfield, face the west, so that, at the resurrection, the dead will rise to face the Son of Man as He comes from out the east with great power and glory.

ORIENTATION

The subject of orientation is an extremely interesting one, and plays a prominent part in many of the customs and practices of the present day. In acknowledgment of the divinity of the Sun the Pagans turned to the east in prayer, and so constructed their temples that even the buildings themselves should pay homage to the rising sun.

We learn from Josephus that as early as Solomon's time the temple at Jerusalem was oriented to the east with great care. It was open to the east, and closed absolutely to the west.

"In plan," says Keary,[1] "it was like an Egyptian temple, the light from the sun at the equinox being free to come along an open passage to reach at last the Holy of Holies. There is evidence too that the entrance of the sunlight on the morning of the spring equinox formed part of the ceremonial; the high priest being in the naos, the worshippers with their backs to the sun could see him by means of the sunlight reflected from the jewels in his garments."

"Temples, with pillars that represented the trees of the sacred groves, had their chief portal almost universally looking toward the east. It is thought that this is due to the fact that the groves, and the temples which represented them, were both indicative of the Garden of Paradise. Again, the portals of Eden where God stationed the Cherubim to keep the way of the tree of life, was on the eastern side of the sacred grove. Not infrequently the approach to these temples was guarded by the figures of the compound sphinx."[2]

The orientation of the Egyptian temples formed

[1] *The Dawn of History*, C. F. Keary.
[2] *The Origin of Pagan Idolatry*, G. S. Faber.

the basis first of the Greek, and later of the Latin temples of worship. Tylor[1] tells us: "It was an Athenian custom for the temple to have its entrance east, looking out through which the divine image stood to behold the rising sun," but for the most part the Greek temples were oriented to the stars that heralded the sunrise, rather than to the orb of day itself.

In India, orientation plays an important part in the daily acts of worship of the Brahman. On rising each day he pays his devotion to the Sun. Standing on one foot, and resting the other against his ankle or heel, he faces the Sun, and stretches out his arms to it. At noon he again worships the Sun, and, sitting with his face to the east, reads his daily portion of the Veda. It is while looking toward the east that his offering of barley and water must be presented to the gods, and in consecrating the sacred fire and sacrificial implements, the east has a holy significance.

On the contrary, the worshippers of Kali, the Death Goddess, the murderous Thugs, regarded the west as sacred, and the Jews, in order that they might not seem to imitate the Pagans in their rites of orientation, placed the sanctuary of their temples toward the west.

It is in Egypt, however, that we find the custom

[1] *Primitive Culture*, Edward B. Tylor.

of orientation most rigidly practised, and a conspicuous feature of temple architecture. The main idea of the builders seems to have been so to arrange them that the chief dates in the year from a solar standpoint should be clearly marked by the orientation of the building, *i.e.*, the solstices and equinoxes.

The temples at Heliopolis and Abydos were unquestionably solstitial temples. The Pyramids of Gizeh were oriented to the equinoxes. Perhaps the most elaborate and important of all the Egyptian solar temples was the magnificent edifice erected at Karnak to the worship of the Sun-God, Amen-Ra. Sir J. Norman Lockyer has given us the following splendid description of this temple.[1] Because it is a typical case of orientation, and one of the most interesting ruins in the world, the author takes the liberty of quoting it in full:

"The solar temple of Amen-Ra at Karnak is the finest Egyptian solar temple which remains open to our examination. It is beyond all question the most majestic ruin in the world. It covers about twice the area covered by St. Peter's at Rome, so that the whole structure was of a vastness absolutely unapproached in the modern ecclesiastical world. It is one of the most soul-inspiring temples which have ever been conceived

[1] *The Dawn of Astronomy*, Sir Norman Lockyer.

The Temple of Amen-Ra, Karnak, Egypt

The Gateway of Ptolemy, Karnak

or built by man. There is a sort of stone avenue in the centre giving a view towards the north-west, and this axis is something like five hundred yards in length. The whole object of the builder of the great temple at Karnak was to preserve that axis absolutely open, the point being that the axis should be absolutely open straight and true. The axis was directed towards the hills on the west side of the Nile where are located the tombs of the kings. From the entrance pylon the temple stretches through various halls of different sizes and details until at last at the extreme end what is called the Sanctuary, Naos, Adytum, or Holy of Holies is reached. The end of the temple at which the pylons are situated is open, the other closed.

"Every part of the temple was built to subserve a special object, viz., to limit the light which fell on its front into a narrow beam and to carry it to the other extremity of the temple into the sanctuary so that once a year when the sun set at the solstice the light is passed without interruption along the whole length of the temple finally illuminating the sanctuary in most resplendent fashion, and striking the sanctuary wall. The wall of the sanctuary opposite to the entrance of the temple was always blocked. The ray of light was narrowed as it progressed inward from the opening toward the sanctuary by a series of doors ingeniously arranged, that acted as the diaphragms of the telescope tube in concentrating the light rays. The reason for this was that the temple was virtually an astronomical observatory, and the idea was to obtain exactly the precise time of the solstice. The longer the beam of light used, the greater is the accuracy that can be obtained. The darker the sanctuary the more obvious will be the patch of light on the end wall, and the more easily can its position be located. It was important to do this two or three days near the solstice in order to get an idea of the exact time at which the solstice took place.

"We find that a narrow beam of sunlight coming through a narrow entrance some five hundred yards away from the door of the Holy of Holies would, provided the temple were properly oriented to the solstice, and provided the solstice occurred at the absolute moment of sunrise, or sunset, according to which the temple was being utilised, practically flash into the sanctuary, and remain there for about a couple of minutes, and then pass away.

"We may conclude that there was some purpose of utility to be served, and the solar temples could have been used undoubtedly among other things for determining the exact length of the solar year. The magnificent burst of light at sunset into the sanctuary would show that a new true solar year was beginning. If the Egyptians wished to use the temple for ceremonial purposes, the magnificent beam of light thrown into the temple at the sunset hour would give them opportunities and even suggestions for so doing; for instance, they might place an image of the god in the sanctuary, and allow the light to flash upon it. We should have a 'manifestation of Ra' with a vengeance, during the brief time the white flood of sunlight fell on it. Be it remembered that in the dry and clear air of Egypt, the sun casts a shadow five seconds after the first little point of it has been seen above the horizon, so that at sunrise and sunset in Egypt the light is very strong, and not tempered as with us."

An extremely interesting feature of Egyptian temple orientation, although it only pertains to temples dedicated to star worship, is found in the fact that the precession of the equinoxes necessitated an alteration of the axis of the temple at long intervals to accommodate the change of direction of the star to which the temple was originally oriented.

At Luxor, and many other places in Egypt, there are to be seen evidences of this change, a later axis having been constructed to meet the new requirements.

In the western world, in ancient times, we find

the same rites of orientation observed and practised as prevailed in the Orient. Among the Sun-worshipping Peruvians the villages were so laid out that they sloped eastward so that the people when they rose each day might behold first of all the deity they worshipped. In the temple of the Sun at Cuzco, the great golden disk that represented the sun was so placed that it received the rising rays of the orb of day, and reflected its light through the edifice erected to its worship.

In ancient Mexico the inhabitants faced the east when they knelt in prayer, as their brother worshippers did in the far east, and though the doors of their temples faced westward, the altar itself was situated in the east. Even the Christianised Pueblo Indians face the east when rising, a survival of their ancient Sun worship. The Sun chief of the Natchez Indians of Louisiana always smoked toward the Sun each morning. The Comanche Indians, when about to take the warpath, present their weapons to the Sun, that their deity may bestow his blessing upon them. The ancient cave temples of the Apalachees of Florida faced eastward, and on festival days the priest waited till the rays of the sun had entered the temple before beginning the ceremonial chants.

According to Tylor,[1] the ceremony of orienta-

[1] *Primitive Culture*, Edward B. Tylor.

tion was unknown in primitive Christianity, but it developed within its first four centuries. It became an accepted custom to turn in prayer toward the east as the auspicious and mystic region of the Light of the World, the Sun of Righteousness, not for the Pagan purpose of adoring the sun itself, and even to-day the boy choristers face the east when they chant the "Gloria."

Orientation played an important part also in the rite of baptism. In ancient times he who was about to embrace the faith faced the west and renounced Satan, with gestures of abhorrence. Then, turning to the east, he acknowledged his faith in Christ, and declared his allegiance to Him. From the fifth century to the time of the Renaissance, the orientation of Christian churches was generally carried out. According to St. John of Damascus, the mystical reasons for this practice were that the crucified Saviour of Mankind faced westward, hence it is fitting that Christians in paying their devotions to Him should face eastward. Again, in the sacred writings, Jesus is called "the east" (*oriens ex alto*) and the hope is expressed that Christians at the last day will see Christ descending in the east. Finally, Christians, when turning to the east during prayer, establish a difference between themselves and the Jews and heretics, for the Jews when praying face west, and certain heretics south, and others north.

In the ninth century there was a strong protest against orientation, but the custom revived later, and to-day all our churches are more or less oriented, particularly those of the Romish and English branches of the Christian Church.

"Any church," says Keary,[1] "that is properly built, will have its axis pointing to the rising of the sun on the Saint's day, i. e., a church dedicated to St. John should not parallel a church dedicated to St. Peter.

"In regard to St. Peter's at Rome, we read that so exactly due east and west was the Basilica, that on the vernal equinox the great doors of the porch of the quadriporticus were thrown open at sunrise, and also the eastern doors of the church itself, and as the sun rose, its rays passed through the outer doors, then through the inner doors, and penetrating straight through the nave, illuminated the High Altar."

There is little doubt that the stones at Stonehenge were so arranged that at sunrise at the summer solstice the shadow of one stone fell exactly on the stone in the centre of the circle, indicating to the priests that the new year had begun. It is thought that fires were lighted on

[1] *The Dawn of History*, C. F. Keary.

this occasion to flash the important news through the country.

Orientation plays an important part in many games and customs in vogue to-day. In the ring games of children, particularly, we see survivals of ancient Sun worship; as, for instance, in the games where the children sing, as they circle round, "Oats, peas, beans, and barley grow," and "Here we go 'round the mulberry bush." The people of early times held that going sunwise was good and lucky, while to move in the contrary direction was inauspicious. The lama monk whirls his praying cylinder in one direction on this account, and fears that some one will turn it contrariwise, in which case it would lose its virtue. These monks also build up heaps of stones in the road, and uniformly pass them on one side as they proceed in one direction, and on the opposite side in returning, in imitation of the sun's circuit.

In India and Ceylon the same circumambulatory customs were practised. As we have seen, it was an old Irish and Scotch custom to go "deazil" or sunwise round houses and graves, and to turn the body in this way at the beginning and end of journeys for luck, as well as at weddings and various ceremonies.

The Irish and Scotch peasants always went westward round a holy well, following the course of the sun, and creeping on their hands and knees, as did the ancient Persians when offering homage at sacred fountains.   In the mystic dances performed at the Baal festivals the gyrations of the dancers were always westward in the track of the sun, for the dance was part of the ancient ritual of Sun worship.   To turn the opposite way, that is against the sun, was considered very unlucky, and was supposed to be an act intimately connected with the purposes of the evil one.   Witches were said to dance that way.

In a religious observance called "paying rounds," much practised by the Irish peasantry when they essayed to cure diseases or bodily ailments, one finds an interesting instance of the custom of going sunwise to produce auspicious results.

It is a general popular belief throughout the United States that in making cake the eggs, or indeed the whole mixture, must be stirred or beaten from beginning to end in the same direction in which the stirring began, or the cake will not be light, and that a custard will curdle if the stirring motion is reversed.   Often it is said that the stirring must be sunwise, the popular expression for this motion being "with the sun."   The same notion is found in Newfoundland and Scotland.

Some matrons in Northern Ohio say that, to

insure good bread, the dough should be stirred with the sun, and that yeast should be made as near sunrise as possible to secure lightness. In New Harbour, Nova Scotia, it is customary in getting off small boats to take pains to start from east to west, and, when the wind will permit, the same custom is observed in getting large schooners under way.

The idea of sunwise movement often appears in the common household treatment of diseases. Before the days of massage, in rubbing for rheumatic or other pains, it was thought best to rub from left to right. It is also said that a corn or wen may be removed by rubbing "with the moon," if by night, and "with the sun," if by day. It is thought that the sun or moon, as the case may be, will draw away all pain and enlargement.

Doubtless a close study of local customs prevailing in different parts of the world will reveal many similar examples and survivals of Sun worship. The subject is an exceedingly interesting one, and reveals above all else the great hold that Sun worship once held on the peoples of the earth.

Chapter XII

Emblematic and Symbolic Forms of the Sun

Chapter XII

Emblematic and Symbolic Forms of the Sun

THERE is much of interest in the study of the symbolic forms of the sun, derived as they are from the mythology and worship of the ancients. Many of the solar symbols enter into designs that embellish works of art of ancient and modern production, but, as symbolism and worship are closely correlated, it is in the study of ecclesiastical architecture, and the structural and artistic adornment of edifices dedicated to worship, that we find a fertile field for tracing these emblems and symbols to their sources.

A knowledge of the origin and true significance of symbolic forms lends interest to many emblems common in the everyday life of the present time, and, in some cases, reveals the curious fact and incongruity of the combination of pagan symbols of worship and anti-pagan ritual. This state of affairs is in evidence in practically all of the modern church edifices. The meaning of these symbols was lost sight of as the wave of Christianity swept onward, and yet so great was the power that these mute forms at one time possessed over men, that, in spite of the decline and the utter extinction of the worship that created them, they continued to live throughout the ages, and many of them in the light of modern times have attained an altered significance, and are as greatly revered and adored to-day as in the ancient days of heathendom.

Although the Canaanite symbol for the sun was an upright stone, the most ancient and popular solar symbol seems to have been the eye. This symbol was naturally suggested by, and in conformity to, the ancient idea that the sun was an all-seeing god, whose penetrating gaze revealed everything that was visible to man. The sun, in short, possessed to primitive minds all the attributes of a great eye gazing down upon the earth.

To the Persians the sun was the eye of Ormuzd. To the Egyptians it was the right eye of Demiurge, and in the *Book of the Dead* the sun is often represented as an eye provided with wings and feet.

To the ancient Hindus the sun was the eye of Varuna. To the Greeks it was the eye of Zeus, and the early Teutons regarded it as the eye of Wuotan, or Woden. Greek mythology, in fact, shows us a race of sun people, the Cyclops, huge giants possessing but one great eye in the centre of their foreheads.

Not only was the sun regarded as a symbol of the eye of the Supreme Being, but in some cases it was thought to represent his full face and countenance.

Vieing in importance with the eye symbol of the sun was the wheel symbol. This was a very ancient conception, and, in this case, the rays of the sun were represented by the spokes of the wheel. As the sun's motion was a matter of great concern to the ancients, its motive power was a subject of much conjecture, and there emerged the fancy that the sun was drawn across the sky by a number of spirited steeds. According to Hindu myth these steeds of the sun were appropriately red or golden in colour.

The symbol of the sun at Sippara was a small circle with four triangular rays, the four angles between being occupied by radiating lines, and the whole circumscribed by a larger circle. The same symbol occurs repeatedly upon the shell gorgets of the ancient Mound-Builders of the western continent.[1]

[1] The Moqui Indian Symbol of the sun is a Greek cross with a small circle in the centre, in which are three marks to indicate the eyes and mouth of a face.

In Greece the symbols of Helios, the Sun-God, were horses' heads, a crown of seven rays, a cornucopia, and a ripened fruit; while the symbols of Apollo were a wolf, swan, stag, dolphin, laurel, and lyre. The ancient Chinese solar symbol was a raven in a circle.

Given the symbol of a wheel, and that of the Sun drawn by dashing steeds, we find in the combination of these symbols the image of the chariot of the Sun, around which so many ancient myths and legends cluster. From the vehicle it was but a step of the imagination to regard the Sun as the charioteer, the supreme deity, driving his flaming car each day across the firmament. The warlike propensities of primitive man were responsible for another very early symbol of the sun, that of a highly burnished shield, while, in a passage of the Persian national epic by Firdusi, the sun is regarded as a golden key which is lost during the night, and the lighting up of the sun each morning was looked upon as an unlocking of the imprisoned orb of day.

Egypt has probably given us more symbols of the sun than any other country. This is doubtless because the Egyptians had a more elaborate form of Sun worship than existed in any other land, in that the different aspects of the sun were, as we have seen, deified. The best-known Egyptian solar symbols were the scarabæus, hawk and globe, lion, and crocodile.

Taking them up in order, the scarabæus or beetle was, according to Pliny, worshipped throughout the greater part of Egypt. It was a symbol particularly sacred to the sun, and is often represented in a boat with wings extended, holding in its claws the globe of the sun. Horapollo claims it was chosen as a solar symbol, owing to the fact that the creature had thirty toes, which equals the number of days in the ordinary solar month. Frequently the claws are represented as clasping a globe, emblematic of the action of the Sun-God Ra at mid-day.

In the great temple at Thebes, a scarab has been recovered with two heads, one of a ram, the symbol of Amen or Ammon, the god of Thebes, the other of the hawk, the symbol of the Sun-God Horus, holding in its claws a globe emblematic of the universe. This scarab has been thought to symbolise the rising sun, and the coming of the spring sun of the vernal equinox in the zodiacal sign of the ram.

The Scarab Beetle

Pliny avers that the claim of the scarab as a solar symbol rests on the fact that in this insect there is some resemblance to the operations of the sun, as one species forms itself into a ball and rolls itself along. The sculptures indicate clearly that the scarabæus represents the orb of the sun. The earliest scarabs date back to about 3900 B.C., and they were regarded as a sacred symbol for a period of over three thousand years. Inasmuch as the scarab was a solar symbol, it was likewise an emblem of immortality, and thus this symbol in its day closely resembled in its true significance the Christian symbol of the cross. The scarab was especially sacred to the Sun-God Amen-Ra, and further symbolised creative and fertilising power. It was the first life appearing after the annual inundation of the river Nile.

Of equal importance to the beetle as a symbol of the sun in Egypt was the hawk, or the hawk and globe, sacred as the emblem of the solar deity. The Sun-God Ra was generally represented as a man with a hawk's head surmounted by a globe or disk of the sun from which an asp issued, and the hawk was particularly known as the type of the sun, worshipped at Heliopolis as the sacred bird, and the representative of the deity of the place.

The winged disk was likewise a solar symbol, and highly regarded by the ancient Egyptians, who considered it an emblem of divine protection. It typified the sun's light and power, transported to the earth on the wings of a bird (possibly the hawk), and the emblem appears on many of the temple walls and over the doorway of numerous dwellings in Egypt.

Porphyry says the hawk was dedicated to the sun, being the symbol of light and spirit, because of the quickness of its motion, and its ascent to the higher regions of the air.   Horapollo thinks that it was chosen as a type of that luminary from its being able to look more intently towards its rays than any other bird, whence also under the form of a hawk they depicted the sun as the Lord of Vision.

Macrobius, Proclus, Horapollo, and others state that the lion was a symbol of the sun, and this is substantiated by the sculptures.   Macrobius claimed further that the Egyptians employed the lion to represent that part of the heavens where the sun was in its greatest force during its annual revolution, the zodiacal sign Leo being called the "abode of the Sun."

The Egyptians, Hindus, Chaldeans, Persians, and Celts all regarded the lion as a solar symbol. Brown[1] tells us: "In the inscriptions of Darya-vush I. at El-Khargeh, the oasis of Ammon, in the Libyan desert, the great god Amen-Ra, the Invisible god revealed in the sun god, is addressed as 'the Lion of the double lions.'   These two lions, two brothers, the two lion gods, are two solar

[1] *The Lion and the Unicorn*, Robert Brown, Jr.

phases, as diurnal and nocturnal, Har and Set, Shu and Tefuut, and as there is but one solar orb, so he is the lion of the double lions. In the funeral ritual the Osirian, or soul seeking divine union and communion with the sun god, prays: 'Let me not be surpassed by the Lion god: Oh, the Lion of the sun, who lifts his arm in the hill [of heaven]' and exclaims: 'I am the Lions, I am the sun. The white lion is the phallus of the sun.'"

The lion and sun form the familiar national standard of Persia, and a Persian coin by Tavernier shows the sun, horned and radiate, rising over the back of a lion.

In many parts of Egypt, in ancient times, the crocodile appears to have been worshipped. This worship was intimately connected with Sun worship, and rested on the analogy between the natural habits of the crocodile and the course of the sun;—as the crocodile spends its days on the land, and at night-fall seeks the water, so the sun, after running its daily course, sinks at evening into the sea. The crocodile, therefore, came to be regarded as a solar symbol, and so figures on the sculptures.

The cat was a conspicuous solar symbol in Egypt. The female of the species was emblematic of Bast or Bubastis, a solar deity, and the male symbolised the great Sun-God Ra.

The Egyptians also represented the sun by the figure of a man sailing in a ship upon the ocean. Sometimes the ship was supported on the back of a crocodile, and again the man appeared floating in the ship, but at the same time seated upon the aquatic lotus, and often the ship was omitted, and the man was supported simply by the lotus. Sometimes the man's place in the calyx of the lotus plant is occupied by the figure of a child, and in the Bembini table, a frog is figured squatting on the floating lotus leaf in place of the man or child.

The Egyptians also represented the sun and moon, Osiris and Isis, as the ox and the cow, and Lady Wilde[1] tells us that these were used at the Irish wake ceremonials until very recently. "There is perhaps," says Faber,[2] "no part of the Gentile world in which the bull and the cow were not highly reverenced, and considered in the light of holy and mysterious symbols. Among the Chinese, the great father Fohi, whose whole history proves him to be the scriptural Noah, is feigned to have had the head of a bull. In the neighbouring empire of Japan, this ancient personage is venerated under the title of 'ox-headed prince of heaven,' and his figure is here again that of a man having the horns of a bull."

---

[1] *Ancient Legends of Ireland*, Lady Wilde.
[2] *The Origin of Pagan Idolatry*, George Stanley Faber.

Bearing in mind the nature and meaning of these solar symbols, the Egyptian sculptures have, for those who study them, a significance that renders them doubly interesting.

The Sun-Gods of the Hindus were represented as seated on the sacred lotus, or floating on the surface of the great deep, either on a leaf, or a huge serpent coiled up in a boatlike form.

In Greece there were legends of the voyage of the solar deity over the ocean, borne in a golden cup, originated, we are told, from the circumstance of the yellow or golden cup of the lotus being employed to represent the ship of the Sun. Indeed, in Hindustan, the cup of the lotus and the ship of the Sun-God Siva mean the same thing. "So strongly," says Faber,[1] "was the idea of a mariner sun impressed upon the minds of the ancient Pagans, that they even transferred it to the sphere. Not content with making the sun sail over the ocean in a ship, they considered the whole solar system as one large vessel in which the seven planets act as sailors, while the sun as the fountain of ethereal light presides as the pilot or captain.

"These eight celestial mariners who navigate the ship of the sphere are clearly the astronomical representatives of the eight great gods of Egypt, all of whom, including the sun as their head, were

[1] *The Origin of Pagan Idolatry*, George Stanley Faber.

wont, according to Porphyry, to be depicted not standing on dry land, but sailing over the ocean in a ship."

The natives of Central America represented the sun by a human head, encircled by diverging rays, and with a great open mouth. This solar symbol was widely spread in all that region. In this representation the tongue is protruding, which signifies that the sun lives and speaks. This is clearly evidenced in the famous Aztec Calendar Stone (Calendario Azteco), also called "Stone of the Sun," which was recovered about the middle of the seventeenth century in the subsoil of the Plaza Major, Mexico City. Terry's *Guide-Book on Mexico* thus describes this interesting relic of antiquity:

"A huge rectangular parallelopipedon of basaltic porphyry, twenty-two feet in diameter, by three feet thick, which weighs twenty-four tons is one of the most interesting of the Aztec relics. . . . This immense specimen, which resembles an irregular mill-stone with a disk carved on it in low relief, evidently served the Aztecs as a calendar stone, and sun-dial. The face is carved with chronological and astronomical signs in geometrical order. The central figure, with a protruding tongue, represents the sun 'Tonatiuh'; the segments radiating toward the edge of the disk are symbolic of its rays.

"Encircling this central figure are seven rings of unequal widths; from the third to the seventh they are incomplete. The inner ring represents two groups of signs, each group containing four symbols. Above the face is an arrowhead, symbolic of the wind, and beneath it a cluster of balls and hieroglyphs. In the rectangles above and below the eagle claws at the right and left are symbols representing the four elements, Air, Fire, Water, and Earth. The symbols on both sides of the upper arrowhead are supposed to represent the years. Five ornamental disks fill the spaces between the symbols. The rectangles of the second ring contain the names of the days of the Aztec month, they begin above the point of the arrowhead and continue toward the left. . . . The third ring contains forty small squares each with five balls supposed to represent days—two hundred in all. Crossing this ring and extending to the sixth are four large arrowheads. The latter ring is the largest of all, and is formed by two huge serpents whose tails terminate in arrowheads ornamented with feathers. The chronological figures between the ends of the tails are thought to correspond to the year 1479 of our era. The human heads ornamented with feathers, eagle claws, disks, ear pendants, and what not represent the gods, (at the left) Tonatiuh, the sun, and (at

The Aztec Calendar Stone

Courtesy of Mrs. J. R. Creelman

Royal Arms of England

the right) Quetzalcoatl, god of the air. The rim
of the huge stone is adorned with conical glyptics,
half stars and balls symbolic of the worlds and
stars. . . . By means of this Calendario the
priests kept their own records, regulated the
festivals and seasons of sacrifices, and made their
astronomical calculations. The symbols show
that they had the means of settling the hours of the
day with precision, the periods of the solstices and
equinoxes, and that of the transit of the sun
across the zenith of Mexico."

Prescott describes this stone in his *Conquest
of Mexico*, and it is now on exhibition in the
National Museum of the City of Mexico.

In the sculptures of the ancient Toltecs, the
Sun- and Moon-Gods are represented by the
symbols of the tiger and the hare respectively.

Akin to the solar symbols of the Canaanites,
the primitive Mexicans erected columns of stone
elaborately carved. These symbolised the sun,
and as Réville puts it, " The sun traces each day the
shadow of these monoliths upon the soil. He
appears to caress and love them, regarding them as
his fellow-workers in measuring time. "[1]

Many of the mystic signs common to pagan
worship are in evidence to-day, "and the High
Churchman decorates the edifice in which he

---

[1] *Native Races of Mexico and Peru*, Albert Réville.

officiates with symbols similar to those which awed
the worshippers of Ashur, Ishtar, or the sun."[1]

Chief among these ecclesiastical solar symbols
is the cross, symbol of the Christian faith, a symbol
that antedated the birth of Christ, and one that
found its origin in solar worship. It occurs upon
the monuments and utensils of every primitive
people, from China to Yucatan. It may be
asked, how did the cross, symbol of the sun, origi-
nate? The following ingenious explanation has
been offered:

"If any one will observe carefully a lamp, or
other bright light, with partially closed eyes, the
answer will be obvious. The rays which appear
to proceed from the luminous point form a cross
of some kind. This is due to the reflection from the
eyelashes, and edges of the eyelids. The evolution of the sun symbol seems to have been as
follows: He was first represented by a circle or
disk as he appears when near the horizon. Observations made when he was shining brightly
revealed the crossed rays. This led to a combination of the circle and cross. If this is correct
the swastika is a modification of the circle and
inscribed cross. Not the least remarkable feature
of this subject is the fact that the most ancient
and universal symbol of the physical sun should

[1] *Ancient Faiths*, Thomas Inman.

for entirely independent reasons continue in use as the sign of the Sun of Righteousness, and the Light of the World."

The simple cross, with perpendicular and transverse arms of equal length, represents the nave and spokes of the solar wheel, sending forth its rays in all directions. In the ancient parish church of Bebington, Cheshire, England, there is to be seen to this day not only the solar wheel, as one of the adornments of the reredos, but deltas, acorns, and Maltese crosses (all of which are pagan symbols) enter profusely into the decorative features of the edifice.

One of the oldest and most widely occurring forms of the cross is the cross with crampons turned to right and left, commonly known as the "swastika," ⊐⌐ the suavastika of India, the Thor's ham ⌐⌐ mer of Western Europe. Professor Max Müller thinks that this symbol represents the vernal sun, and that it is an emblem of life, health, and creative energy. It is thought to have arisen from the conception of the sun as a rolling wheel.

The halo depicted as encircling the heads of the saints, and those endowed with holy attributes, is clearly a solar symbol, and the wheel symbol suggested by the disk of the sun was often used as an emblem of God.

In the chancel of a church in New England to-day we see in the mural decorations symbols that typify the ancient deification of the Sun, and originated from that worship, such as the disk fringed with darting rays, the sun symbol, in the centre of which is the Christ name symbol, a strange and incongruous combining of the symbols of antagonistic and widely differing cults.

In another church in the same locality is the symbol of the six-pointed star, enclosing the all-seeing eye. This double equilateral triangle is one of the most sacred of all the emblems of Pythagoras, and was revered for ages as the seal of King Solomon. It is also an important Masonic emblem.

The strange part of this study of symbolism is that the significance of these heathen emblems should be utterly meaningless to the multitudes who worship in their sight, which indicates an indifference to a knowledge of symbolism not in accord with the desire oftentimes emphasised for it, and the great number of emblems which embellish and adorn modern ecclesiastical edifices.

The emblems of heraldry perpetuate the symbolism expressive of the solar worship of primitive times. We see the Royal Arms of England, supported by the solar lion and the lunar unicorn.

"These two creatures," says Brown, "are naturally antagonistic. In the ancient myth, the Unicorn, when rushing at the Lion, sticks his horn fast in a mythic Tree, behind which his opponent has taken refuge, and the Lion coming round devours him whilst thus defenceless. This is the explanation of the myth. The Lion-Sun flies from the rising Unicorn-Moon, and hides behind the Tree or Grove of the Under-world, the Moon pursues, and sinking in his turn, is caught in this mysterious Tree, and sunslain."[1]

In many escutcheons are to be seen solar symbols already referred to, as, for instance, in the cut, the escutcheon of a Greek-letter fraternity shows the winged sun disk and the all-seeing eye.

In astrology also, solar symbolism plays an important part, such as "the rules which connect the sun with gold, with heliotrope and pæony, with the cock which heralds day, and with magnanimous animals such as the lion and bull."

There is to be found in certain old brick houses in England a curious solar symbol. It consists of a flat piece of iron five or six inches in length, shaped somewhat like the letter "S," which was placed upon the house walls about the line of division between the

[1] *The Unicorn*, Robert Brown, Jr.

first and second stories. It is still used in Herefordshire. There, it is said that these irons are in the nature of talismans, and are supposed to protect the house from fire and collapse.

Brown tells us that "Masonic tradition is but one of the numerous ancient allegories of the yearly passage of the personified sun among the twelve constellations of the zodiac, being founded on a system of astronomical symbols and emblems, employed to teach the great truths of omnipotent God and immortality." [1] Its symbolism, therefore, is closely associated with solar symbolism and interesting to note in this connection. The word "Masonry" is said to be derived from a Greek word which signifies "I am in the midst of heaven," alluding to the sun. Others derive it from the ancient Egyptian "Phre," the sun, and "Mas," a child, Phre-mas, *i.e.*, children of the sun, or sons of light. From this we get our word "Freemason."

Masons are instructed to travel eastward in search of light, as the sun rises in the east. The initiation into all the ancient mysteries was a drama founded on the astronomical allegory of the death and resurrection of the Sun, and impressed on the mind of the candidate the unity of God and the immortality of man. These facts are taught in the ritual of the Third Degree.

[1] *Stellar Theology*, Robert Brown, Jr.

Greek-Letter Fraternity Escutcheon

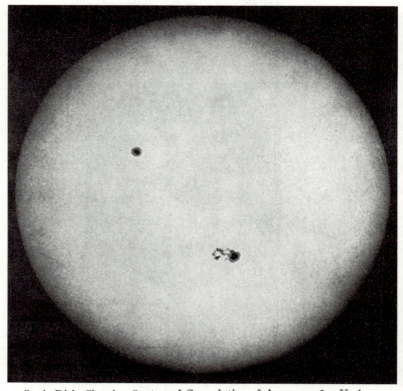

Sun's Disk, Showing Spots and Granulation, July 30, 1906.   Yerkes

The Sun, overwhelmed by the three autumn months, returns to life at the vernal equinox, and is exalted at the summer solstice. In this drama the candidate was required to represent the Sun, and a solar significance characterises the whole ritual.

The following references to the symbolism of Masonry are taken from *Stellar Theology* by Robert Brown, Jr.

"The Lodge should be situated East and West, because the Sun, the glory of the Lord, rises in the East and sets in the West. A lodge has three lights situated in the East, West, and South. The Master's place is in the East, whence the sun rises, the Senior Warden's in the South, the point the sun occupies at mid-day.

"E.A.M. signifies the sun, F.C.M. the moon, and M.M. the sun, Benevolent God of Fire.

"O.G.M.H.A. is derived from two roots signifying the origin or manifestation of light, also he who was, and is. The source of eternal light, the sun taken as an emblem of Deity.

"O.G.M.H.A. represents the sun. The three steps delineated on the master's carpet have an obvious reference to the three steps or degrees by which the initiated becomes a Master-mason. They allude to the constellations Taurus, Gemini, and Cancer (emblematic of three steps), by means of which the sun ascends to the summit of the Royal Arch.

"The emblem of the Blazing Star alludes to the Sun as a symbol of Deity. The rite of Circum-ambulation has a direct solar allusion, as it was always performed from right to left, in imitation of the apparent course of the sun from East to West by way of the South.

"Masons celebrate June 24th and December 27th. These dates have a purely astronomic significance, and refer to the summer and winter solstice, the periods of great festivals and celebrations throughout the ancient world.

"The symbol of the all-seeing eye is distinctly solar in its character. In most of the ancient languages of Asia, eye and sun are expressed by the same word. In like manner Masons have emblematically represented the omniscience of the great Architect of the Universe.

"The significance of the Pillars of the Porch is of interest. In every Lodge there are two pillars surmounted by globes. These represent the pillars in the porch of King Solomon's temple. The Egyptian temples always contained two such pillars, one called 'Boaz,' meaning the sun, on the north side, the other 'Jachin,' referring to the moon, on the south.

"The corner-stone of the Lodge is placed in the

North-East, as the sun on June 21st rises in the North-East."

According to Professor Worsaal, the ring cross is a symbol of the sun, and belongs to the later stone age of Scandinavia. It was also the Chaldean solar symbol. The same writer places amongst the emblems of the later bronze age the symbol of the wheel cross, which is considered a symbol of the sun.

The subject of solar symbolism has been only briefly touched upon in the foregoing, and a close study of its many features affords a rich field for research that should prove of fascinating interest to scholars and antiquarians alike.

Chapter XIII

The Sun Revealed by Science

Chapter XIII

The Sun Revealed by Science

INASMUCH as the foregoing chapters have treated of the mythology of the sun, its worship, and the curious legends and traditions regarding that luminary, any discussion of the sun astronomically speaking, and considered in the light of modern science, would seem out of place and irrelevant. Still, the fact that lore does not necessarily embrace the archaic and mythological alone but may include fact as well as fancy, renders a reference to what scientific solar research has accomplished entirely proper in this volume.

Further than this, such matter must also serve to bring to mind what man has achieved in his desire to know more about the great system of which the world is such an insignificant part.

The pages of the history of man that record his knowledge of the sun, as revealed by long years of research, are among the most interesting, if not the most wonderful, that human progress has inscribed. Marvellous strides have been made

toward solving the many enigmatic problems that
the subject presents, and man has of late, through
the wonderful ingenuity of those tireless and gifted
toilers in this field, accomplished so much, that a
résumé of what has been done even in the last ten
years reads like the pages of a romance.

Our knowledge of the sun, physically and scienti-
fically speaking, is of comparatively recent date.
In fact, only a hundred years ago the question of
the habitability of the sun was discussed in all
seriousness by the astronomers of the day, but
with the invention of new apparatus, the construction of powerful telescopes, the development of
spectroscopic research, and the zeal of intelligent
men, this question and many other difficult solar
problems have been positively solved, and no
scientist of standing believes to-day that the sun
is capable of sustaining life as we know it.

Even half a century ago our scientific knowledge
of the sun might easily have been contained in a
brief essay, but, so rapid has been the advance,
that to-day two large volumes can hardly do the
subject justice. An apology is therefore due the
reader for this attempt to treat in brief such a
vast subject. No attempt will be made to discuss
technicalities, and it will suffice to state briefly the
facts that have rewarded the efforts of man in his
endeavour to wrest from that great and distant

power-house of vitality the secrets that have baffled the ages.

Looking back on the centuries shrouded in ignorance and superstition, when imagination was a more potent factor than study and investigation, when myth and legend were of more concern than fact and knowledge, we realize what an age of progress this is, and how the rigid and scientific methods of to-day have won their way through the superstitious mists of past ages, affording us a glimpse of the marvels of the universe, and the beauties of a nature sublime in its perfection.

Perhaps the first and most important fact to mention in discussing the sun from a physical standpoint, is that the sun is a star, the nearest star to the earth, so that the more we know about the sun the greater will be our knowledge of the stars. It is a fact, therefore, that we know no light save starlight, for the luminosity attributed to some of the planets, and the transient comets that come our way, is too infinitesimal to be considered.

Apart from a knowledge of what the sun really is, the chief consideration is its distance from the earth and its size.

There are several reliable and totally independent methods of calculating the distance that separates the earth from the sun. They are derived from the following investigations as given in *The Sun* by Professor Charles G. Abbot:

1. Heliometer work on the minor planets.
2. Measurements involving the asteroid Eros.
3. Gravitational methods.
4. Observations of the eclipses of Jupiter's satellites.
5. From the constant of aberration of light.

A mean of the results of investigations, according to these methods, yields 92,930,000 miles as the sun's distance from the earth. This may be regarded as very close, comparatively speaking, to the actual distance.

Given the distance to the sun, it is a comparatively simple calculation to state its size. Professor Abbot's estimate of the sun's diameter is 865,000 miles. This is about 110 times the diameter of the earth, and the volume or bulk of the sun is calculated to be more than 1,306,000 times that of the earth.

In spite of the sun's enormous bulk it has been weighed, and it is found that the sun's mass is about 332,800 times that of the earth. The earth's weight is estimated to be six thousand millions of millions of tons. If a human being were transported to the sun, he would weigh two tons, a weight sufficient to crush him to death, the force of gravity at the sun's surface being twenty-seven times that of the earth.

We find that this enormous body, like the earth, turns on an axis passing through its centre, and inclined about seven degrees to the plane of the ecliptic.   As the sun is not a solid body, at least at its surface, it has been proved that at various latitudes the speed of axial rotation varies.   "The sidereal rotation of the average solar surface is about 24.6 days at the equator, 26.3 days at $\pm$ 30° latitude, 31.2 days at $\pm$ 60°, and 35.3 days at $\pm$ 80." [1]   At the sun's equator, therefore, the speed of axial rotation is more than a mile a second.

Although it is interesting, and a tribute to man's intelligence, to know the sun's distance from us, its size, and period of rotation, the importance of the sun to man lies in the fact that it sustains life on the earth by sending to it a sufficient supply of heat and light.   Without this portion of the sun's bounty, man would soon cease to exist.   Consequently it is of the greatest importance that we should study the sun with a view of ascertaining the nature, quantity, and enduring qualities of these essential elements.

A careful study of the sun's surface has revealed the fact that the sun is not a solid body as we have

[1] *The Sun*, Professor Charles G. Abbot.

remarked, but composed of a series of enveloping layers pressing one upon the other, made up of matter wholly gaseous or vaporous. Some authorities claim that liquids and even solids exist in the sun, but it is not for us to discuss the merits of these contentions, or set forth the technical arguments which favour these views.

The surface of the sun which we see when we view the sun through a smoked glass is called the photosphere. In the telescope this enveloping layer has a granulated appearance. It looks as if it were sprinkled over with grains of rice, and the projected telescopic image of the sun will reveal to any one that the centre of the sun's disk is brighter than its limb. This proves that the sun is surrounded by an atmosphere that absorbs to a certain extent the light that emanates from it.

The photosphere is merely the shell of the sun, several hundred miles deep, and probably gaseous in its nature. Above the photosphere is a layer of gases called the chromosphere, which can be observed by the spectroscope, and by direct vision during total eclipses.

Still above the chromosphere there is a layer of thin gases called the corona, "a soft silver radiance," the beauties of which a total eclipse alone reveals.

The nature of the corona has long presented a

baffling problem to astronomers. Its shape varies in accordance with the eleven-year sun-spot period, and it seems electrical or magnetic in its nature. It streams out from the sun many millions of miles until its rays are lost in the rare regions of interstellar space.

These glowing layers of gases thrust out great waves of light and heat to the very confines of the solar system, and we know not how much farther. We do know that the planet Neptune receives its light from the sun, and this planet, now the outpost of our system, circles the sun at a distance of 2,792,000,000 miles from it.

"Of the potency of the sun's rays we can form but a feeble conception, for the amount received by the earth is, it has been calculated, but one twenty-two-hundred millionth of the whole. Our annual share would, it is supposed, be sufficient to melt a layer of ice spread uniformly over the earth to a depth of one hundred feet, or to heat an ocean of fresh water sixty feet deep from freezing to boiling point. The illuminating power of the sun has to be expressed in language of similar profundity. Thus it has been calculated to equal that which would be afforded by 5563 wax candles concentrated at a distance of one foot from the observer. Again it has been concluded that no fewer than one half a million of full moons shining all at once would be required to make up a mass of light equal to that of the sun.'"[1]

Newcomb, writing of the stupendous forces set in play merely in the outer portions of the sun's globe, writes: "Perhaps the explosion of powder when a thirteen-inch cannon is fired is as striking an example of the force of ignited gases as we are familiar with. Now suppose every foot of space in a whole county covered with such cannon, all pointed upward, and all being discharged at once. The result would compare with what is going on inside the photosphere, about as a boy's pop-gun compares with the cannon."[2]

The average temperature of the photosphere, according to Professor Abbot, is about 6200° and possibly near 7000° absolute centigrade.

It is quite impossible to form any comprehension of the terrific expenditure of heat and light from the sun, and the first question that arises in this connection, and the one that is absolutely vital to our interest is, how are the sun fires maintained? How can this enormous waste go on, day in day out, and not diminish, so as to render our globe uninhabitable in a short time?

It has been proved that if the sun were of the nature of a white-hot ball it would cool off so

---

[1] *The Story of the Solar System*, G. F. Chambers.
[2] *Astronomy for Everybody*, Simon Newcomb.

rapidly that its heat could not last more than a few centuries.   How then is the sun's heat replenished? Science thus makes answer: As the sun gives out its store of heat it contracts, and contraction itself produces heat.   The streams of matter projected upward from the interior of the sun become cooled as they reach the surface and fall back.   The heat generated by the fall of this matter, a process eternally going on, replenishes the sun fires, and thus offsets the wasted energy of the sun.

It has been found that in order to keep pace with the sun's output of heat, it is only necessary for the sun's diameter to contract about a mile in twenty-five years, or four miles a century, a shrinkage that would not be perceptible for thousands of years.   Although the sun is inevitably doomed to extinction, and man shall cease to inhabit the earth, the time is so distant that it must be reckoned in millions of years.

This theory of a steadily shrinking sun implies, of course, that millions of years ago the sun was much larger than it is to-day, and so great is the sun supposed to have been that at one time it is thought it occupied the space now filled by the entire solar system.   It was then nothing but a huge nebula.   This theory is the basis of the celebrated Nebular Hypothesis, which is offered as an explanation of the origin of the solar system.

According to this hypothesis, the solar system is the result of the contraction of a nebula in the course of millions of years. This hypothesis is, however, by no means wholly accepted by the scientific world to-day.

We come now to a brief account of solar phenomena, as revealed by instruments of precision in the hands of skilled men, who have devoted their lives to solar research. Probably the best known solar phenomenon is the appearance from time to time on the sun's disk of sun-spots. They are dark and irregular in shape, and range in size from minute points scarcely visible in the largest telescope, to naked-eye spots thousands of miles in extent, that would completely engulf the entire earth if it were placed in one. For instance, the spot of November 16, 1882, had an area of over two billion square miles.

The spots are composed of a dark nucleus or central core surrounded by a fringe a trifle lighter in hue called the umbra, and this in turn is surrounded by a border known as the penumbra. The spots, strangely enough, are found for the most part in a zone extending thirty-five degrees on each side of the sun's equator, although they are seldom seen directly under the equator. The latitude of greatest frequency is seventeen or eighteen degrees. The sun's rotation carries the spots across the sun's disk, and changes in their extent and shape can be easily observed. Frequently they endure while invisible to us, and on their reappearance exhibit marked changes in size and contour. It has been proved that some of the sun-spots have an independent motion of their own, and they last from a few days to a month.

The ancients noticed especially large spots when the sun was near the horizon, but took them either for planets in conjunction with the sun, or a phenomenon which they could not explain. There are Chinese records of forty-five sun-spots having been seen in the interval 301 to 1205 A.D. Adelmus, a Benedictine monk, observed a black spot on the sun March 17, 807, and Fabricius in December, 1610, was able to follow a conspicuously large sun-spot for some time. Nowadays the sun is photographed at some one of our observatories every clear day, and a faithful record of the number and location of every spot on the sun's surface is kept for purposes of reference and discussion.

Perhaps the most interesting fact concerning sun-spots is the law that relates to their periodicity. To the perseverance and unremitting observations of the sun by Schwabe of Dessau from 1826 to 1868 is due the establishment of this law. His records showed that every eleven years marked a period of great sun-spot activity, just as an eleven-year period marked an epoch of sun-spot quiescence. The future years that mark these periods are given below for the next fifty years:

Maximum Number	Minimum Number
1916	1922
1927	1933
1938	1944
1949	1955
1960	1966

In 1908 Dr. Hale of the Mount Wilson Solar Observatory, one of the most eminent investigators in solar research work, discovered the existence of a magnetic field in sun-spots. This confirmed a fact previously noted that terrestrial magnetic disturbances and displays of the aurora borealis frequently accompanied the appearance of a large sun-spot. Recent research has revealed the fact that "the spots are vortices similar in form to ocean waterspouts. The stem of the vortex is the umbra of the sun-spot, the spreading top the penumbra." [1] They savour of the nature of great cyclones, circling with wonderful rapidity, a very maelstrom of whirling matter evolving currents of electricity that produce a magnetic field. It would appear that these stupendous convulsions on the

[1] *The Sun*, Charles G. Abbot.

sun's surface are of sufficient power to penetrate our atmosphere and produce magnetic and electrical phenomena.   Space does not permit of further discussion of this subject which in interest ranks first of the solar phenomena.

Closely allied to the phenomena of sun-spots are the so-called "faculæ," collections of small spots which are brighter than the photosphere.   These areas of disturbance are revealed by the spectroheliograph, a wonderful instrument invented by Professor George E. Hale, which enables us to photograph the sun by the light of a single ray of the spectrum.   When the sun is photographed by calcium light, faculæ are found on every part of the sun, which proves them to be evidences of eruptions of gases chiefly composed of calcium.

Beyond question, the most spectacular phenomena visible in connection with the sun, excepting a total eclipse, are the solar prominences as they are called, great tongues of hydrogen flame that are projected outward from the sun's limb thousands of miles into space, at a speed sometimes of hundreds of miles a second.   These wonderful flames are of such extent that the earth if enveloped by them would be "as a grain of sand in the flame of a candle."   The spectroscope and spectroheliograph enable the observer to command a view of the sun's limb at will, and at times of sun-spot activity there are often marvellous displays of the prominences. An extraordinary outburst was witnessed on September 7, 1871, by the late Professor Young. At noon he observed a prominence that had remained unchanged since noon of the day previous, a long low quiet-looking cloud, not very dense or brilliant, or in any way specially remarkable for its size. In a half-hour interval this great flame sprang into activity and was shattered to pieces. The solar atmosphere "was filled with flying débris" and some of these portions reached a height of one hundred thousand miles above the solar surface, moving at a velocity which even at the distance of ninety-three million miles was almost perceptible to the eye. These fragments doubled their height in ten minutes. Another remarkable outburst was witnessed by Professor Tacchini of Rome, January 30, 1885, when the flame reached a height of one hundred and forty-two thousand miles.

As the sun is the nearest star, a comparison of its light with the bright stars is of interest. Stellar brightness is expressed in magnitudes, a star of the first magnitude being about 2.5 times brighter than a second-magnitude star. Sirius, the brightest star in the heavens, has a magnitude of − 1.4, and Aldebaran, the ruddy star in the eye of the Bull, has a magnitude of 1.1. The sun's magnitude is calculated to be − 26.5 and gives the earth ninety

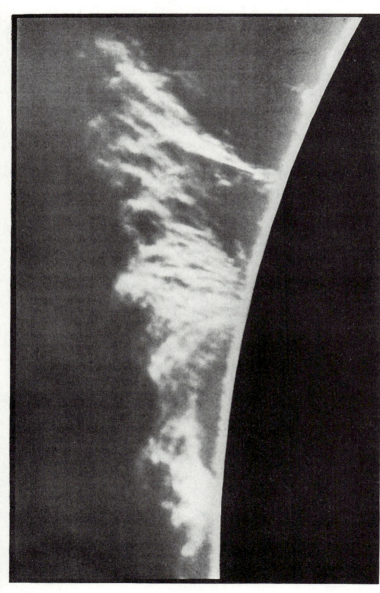

Photo at Mt. Wilson Solar Observatory

Sun's Limb, Showing Prominences 80,000 Miles High, August 21, 1909

billion times the light of Aldebaran. So remote are the stars that, if the sun occupied the place of Aldebaran, it would appear as a faint star of the fifth magnitude.

One of the most interesting features connected with the sun is its motion among the stars. We are inclined to think of the sun as motionless in space, and as merely revolving on its axis. It has been proved, however, that the stars in a particular region of the heavens are closing in toward each other, much as trees widely separated would appear to crowd together as we recede from them. The stars in the opposite part of the sky, on the contrary, appear to be widening out; that is, the distance relatively separating them seems to be increasing. This clearly indicates that we are moving toward these stars, drawn along by the sun, which is pursuing a journey through space in an apparently straight line. Its goal appears to be the region between the constellations Hercules and Lyra, not far distant from the brilliant first-magnitude star Vega. The rate of speed maintained by the sun is estimated from ten to twelve miles a second.

Professor Abbot gives as the apex of the solar system, this goal of the sun's journey, as R.A. 270°, Declination + 30°.

There remains to record the elements that have

been found to exist in the sun. The better known of these are: Very strong, hydrogen, helium, magnesium, calcium. Strong, manganese, iron, barium. Not very strong, carbon, aluminum, vanadium. Very weak, nickel, cobalt, lead. Possibly shown, zinc, tantalum. Doubtful, silicon.[1]

An exceedingly important problem in connection with the sun is now being attacked; namely, the question as to the variability of its light. If the sun can be proved to be a variable star, and its period and range determined, a wonderful advance will have been made in solar physics. There seems every reason to believe now that there is a slight variation in sunlight, but sufficient data are lacking at present for a positive determination as to the range of this variation. When this fact is once fully established we shall be in a position to apply the knowledge to problems of meteorology, and it may be possible to make long-period weather forecasts remarkable for their reliability, an inestimable boon to the race.

The author must again apologise for this extremely brief and fragmentary account of our scientific knowledge of the sun. The subject, however, is highly technical, and extended discussion is irrelevant in a popular treatise of this nature. An attempt has been made to show that

[1] *The Sun*, Charles G. Abbot.

man's ingenuity and intelligence have enabled him to cope with, and in many cases to solve, the problems that the solar phenomena present, and the scientific knowledge of the sun extant to-day, compared with the superstition and ignorance concerning it in the past, is the best evidence of man's intellectual advance.

Bibliography

THE SUN	Charles G. Abbot
NORSE MYTHOLOGY	. R. B. Anderson
THINGS CHINESE	. J. Dyer Ball
THE MYTHOLOGY AND FABLES OF THE ANCIENTS	The Abbé Banier
MYTHS OF ASTRONOMY	J. F. Blake
BUSHMAN FOLK-LORE . W. H. I. Bleek and L. C. Lloyd
AMERICAN ANTIQUITIES .	Alexander W. Bradford
OBSERVATIONS ON THE POPULAR ANTIQUITIES OF GREAT BRITAIN .	. John Brand
THE UNICORN . . . . ⎱ STELLAR THEOLOGY . . . ⎰	Robert Brown, Jr.
A NEW SYSTEM OF ANCIENT MYTHOLOGY .	Jacob Bryant
HINDU ASTRONOMY	. W. Brennand
THE MYTHS OF THE NEW WORLD .	Daniel G. Brinton
NURSERY TALES AND TRADITIONS OF THE ZULUS	. Rev. Canon Callaway
THE MYTHOLOGY OF THE ARYAN NATIONS	Sir George W. Cox
A CHILD'S GUIDE TO MYTHOLOGY . ⎱ ANCIENT MYTHS IN MODERN POETS ⎰	Helen A. Clarke
CREATION MYTHS OF PRIMITIVE AMERICA	Jeremiah Curtin
OLD ENGLISH CUSTOMS . .	. P. H. Ditchfield
THE ORIGIN OF PAGAN IDOLATRY .	. George S. Faber
MYTHS AND MYTH MAKERS . . .	John Fiske
ASTRONOMICAL MYTHS . . .	Camille Flammarion
LAOS FOLK-LORE OF FARTHER INDIA	Katherine N. Fleeson
THE GOLDEN BOUGH	J. G. Frazer
MYTHOLOGY AMONG THE HEBREWS .	. Ignaz Goldhizer

CURIOUS MYTHS OF THE MIDDLE AGES .   Baring-Gould
MYTHS AND SONGS FROM THE SOUTH
    PACIFIC . . . .   Rev. W. W. Gill
POLYNESIAN MYTHOLOGY . .   Sir George Gray
ZOÖLOGICAL MYTHOLOGY .   Angelo de Gubernatis
GREEK LIFE . . . .   John M. Hall
HISTORY OF MAN .   Hon. Henry Home (of Kames)
NOTES ON THE FOLK-LORE OF THE
    NORTHERN COUNTIES OF ENGLAND   William Henderson
POPULAR ROMANCES OF THE WEST OF
    ENGLAND . . . .   Robert Hunt
ANCIENT FAITHS . . . .   Thomas Inman
THE DAWN OF HISTORY . . ⎫
CURIOSITIES OF FOLK-LORE . . ⎬   Charles F. Keary
OUTLINES OF PRIMITIVE BELIEF . ⎭
CURIOSITIES OF INDO-EUROPEAN
    TRADITION AND FOLK-LORE .   Walter K. Kelley
BABYLONIAN RELIGION AND MYTHOLOGY .   L. W. King
CUSTOM AND MYTH . . .   Andrew Lang
THE DAWN OF ASTRONOMY .   Sir J. Norman Lockyer
THE DAWN OF THE WORLD .   C. Hart Merriam
CHIPS FROM A GERMAN WORKSHOP .   Max Müller
MANUAL OF MYTHOLOGY .   Alexander S. Murray
ARTICLE IN DUBLIN REVIEW, VOL. XXXIII   Frederic A. Paley
ROUND THE YEAR WITH THE STARS . ⎫
HALF HOURS WITH THE SUMMER STARS ⎭   Mary Proctor
MYTHS AND MARVELS OF ASTRONOMY   Richard A. Proctor
RUSSIAN FOLK-TALES . .   W. R. S. Ralston
THE NATIVE RELIGIONS OF MEXICO AND
    PERU . . . .   Albert Réville
PERUVIAN ANTIQUITIES .   Rivero and von Tschudi
NORTHERN MYTHOLOGY . .   Benjamin Thorpe
HISTORY OF THE EGYPTIAN RELIGION .   Dr. C. P. Tiele
RESEARCHES IN THE EARLY HISTORY ⎫
    OF MANKIND . . . . ⎬   Edward B. Tylor
PRIMITIVE CULTURE . . . ⎮
ANTHROPOLOGY . . . . ⎭

ANCIENT LEGENDS OF IRELAND . . Lady Wilde
THE MANNERS AND CUSTOMS OF THE
   ANCIENT EGYPTIANS . Sir J. Gardner Wilkinson
JOURNAL OF THE POLYNESIAN SOCIETY
JOURNAL OF AMERICAN FOLK-LORE
REPORTS OF THE U. S. BUREAU OF ETHNOLOGY

Index

A CATALOG OF SELECTED DOVER
BOOKS IN ALL FIELDS OF INTEREST

CONCERNING THE SPIRITUAL IN ART, Wassily Kandinsky. Pioneering work by father of abstract art. Thoughts on color theory, nature of art. Analysis of earlier masters. 12 illustrations. 80pp. of text. 5⅜ x 8½.   0-486-23411-8

CELTIC ART: The Methods of Construction, George Bain. Simple geometric techniques for making Celtic interlacements, spirals, Kells-type initials, animals, humans, etc. Over 500 illustrations. 160pp. 9 x 12. (Available in U.S. only.)   0-486-22923-8

AN ATLAS OF ANATOMY FOR ARTISTS, Fritz Schider. Most thorough reference work on art anatomy in the world. Hundreds of illustrations, including selections from works by Vesalius, Leonardo, Goya, Ingres, Michelangelo, others. 593 illustrations. 192pp. 7⅛ x 10¼.   0-486-20241-0

CELTIC HAND STROKE-BY-STROKE (Irish Half-Uncial from "The Book of Kells"): An Arthur Baker Calligraphy Manual, Arthur Baker. Complete guide to creating each letter of the alphabet in distinctive Celtic manner. Covers hand position, strokes, pens, inks, paper, more. Illustrated. 48pp. 8¼ x 11.   0-486-24336-2

EASY ORIGAMI, John Montroll. Charming collection of 32 projects (hat, cup, pelican, piano, swan, many more) specially designed for the novice origami hobbyist. Clearly illustrated easy-to-follow instructions insure that even beginning papercrafters will achieve successful results. 48pp. 8¼ x 11.   0-486-27298-2

BLOOMINGDALE'S ILLUSTRATED 1886 CATALOG: Fashions, Dry Goods and Housewares, Bloomingdale Brothers. Famed merchants' extremely rare catalog depicting about 1,700 products: clothing, housewares, firearms, dry goods, jewelry, more. Invaluable for dating, identifying vintage items. Also, copyright-free graphics for artists, designers. Co-published with Henry Ford Museum & Greenfield Village. 160pp. 8¼ x 11.   0-486-25780-0

THE ART OF WORLDLY WISDOM, Baltasar Gracian. "Think with the few and speak with the many," "Friends are a second existence," and "Be able to forget" are among this 1637 volume's 300 pithy maxims. A perfect source of mental and spiritual refreshment, it can be opened at random and appreciated either in brief or at length. 128pp. 5⅜ x 8½.   0-486-44034-6

JOHNSON'S DICTIONARY: A Modern Selection, Samuel Johnson (E. L. McAdam and George Milne, eds.). This modern version reduces the original 1755 edition's 2,300 pages of definitions and literary examples to a more manageable length, retaining the verbal pleasure and historical curiosity of the original. 480pp. 5³⁄₁₆ x 8¼.   0-486-44089-3

ADVENTURES OF HUCKLEBERRY FINN, Mark Twain, Illustrated by E. W. Kemble. A work of eternal richness and complexity, a source of ongoing critical debate, and a literary landmark, Twain's 1885 masterpiece about a barefoot boy's journey of self-discovery has enthralled readers around the world. This handsome clothbound reproduction of the first edition features all 174 of the original black-and-white illustrations. 368pp. 5⅜ x 8½.   0-486-44322-1

FRENCH STORIES/CONTES FRANÇAIS: A Dual-Language Book, Wallace Fowlie. Ten stories by French masters, Voltaire to Camus: "Micromegas" by Voltaire; "The Atheist's Mass" by Balzac; "Minuet" by de Maupassant; "The Guest" by Camus, six more. Excellent English translations on facing pages. Also French-English vocabulary list, exercises, more. 352pp. 5⅜ x 8½. 0-486-26443-2

CHICAGO AT THE TURN OF THE CENTURY IN PHOTOGRAPHS: 122 Historic Views from the Collections of the Chicago Historical Society, Larry A. Viskochil. Rare large-format prints offer detailed views of City Hall, State Street, the Loop, Hull House, Union Station, many other landmarks, circa 1904-1913. Introduction. Captions. Maps. 144pp. 9⅜ x 12¼. 0-486-24656-6

OLD BROOKLYN IN EARLY PHOTOGRAPHS, 1865-1929, William Lee Younger. Luna Park, Gravesend race track, construction of Grand Army Plaza, moving of Hotel Brighton, etc. 157 previously unpublished photographs. 165pp. 8⅞ x 11¾. 0-486-23587-4

THE MYTHS OF THE NORTH AMERICAN INDIANS, Lewis Spence. Rich anthology of the myths and legends of the Algonquins, Iroquois, Pawnees and Sioux, prefaced by an extensive historical and ethnological commentary. 36 illustrations. 480pp. 5⅜ x 8½. 0-486-25967-6

AN ENCYCLOPEDIA OF BATTLES: Accounts of Over 1,560 Battles from 1479 B.C. to the Present, David Eggenberger. Essential details of every major battle in recorded history from the first battle of Megiddo in 1479 B.C. to Grenada in 1984. List of Battle Maps. New Appendix covering the years 1967-1984. Index. 99 illustrations. 544pp. 6½ x 9¼. 0-486-24913-1

SAILING ALONE AROUND THE WORLD, Captain Joshua Slocum. First man to sail around the world, alone, in small boat. One of great feats of seamanship told in delightful manner. 67 illustrations. 294pp. 5⅜ x 8½. 0-486-20326-3

ANARCHISM AND OTHER ESSAYS, Emma Goldman. Powerful, penetrating, prophetic essays on direct action, role of minorities, prison reform, puritan hypocrisy, violence, etc. 271pp. 5⅜ x 8½. 0-486-22484-8

MYTHS OF THE HINDUS AND BUDDHISTS, Ananda K. Coomaraswamy and Sister Nivedita. Great stories of the epics; deeds of Krishna, Shiva, taken from puranas, Vedas, folk tales; etc. 32 illustrations. 400pp. 5⅜ x 8½. 0-486-21759-0

MY BONDAGE AND MY FREEDOM, Frederick Douglass. Born a slave, Douglass became outspoken force in antislavery movement. The best of Douglass' autobiographies. Graphic description of slave life. 464pp. 5⅜ x 8½. 0-486-22457-0

FOLLOWING THE EQUATOR: A Journey Around the World, Mark Twain. Fascinating humorous account of 1897 voyage to Hawaii, Australia, India, New Zealand, etc. Ironic, bemused reports on peoples, customs, climate, flora and fauna, politics, much more. 197 illustrations. 720pp. 5⅜ x 8½. 0-486-26113-1

THE PEOPLE CALLED SHAKERS, Edward D. Andrews. Definitive study of Shakers: origins, beliefs, practices, dances, social organization, furniture and crafts, etc. 33 illustrations. 351pp. 5⅜ x 8½. 0-486-21081-2

THE MYTHS OF GREECE AND ROME, H. A. Guerber. A classic of mythology, generously illustrated, long prized for its simple, graphic, accurate retelling of the principal myths of Greece and Rome, and for its commentary on their origins and significance. With 64 illustrations by Michelangelo, Raphael, Titian, Rubens, Canova, Bernini and others. 480pp. 5⅜ x 8½. 0-486-27584-1

THE MALLEUS MALEFICARUM OF KRAMER AND SPRENGER, translated by Montague Summers. Full text of most important witchhunter's "bible," used by both Catholics and Protestants. 278pp. 6⅝ x 10. 0-486-22802-9

SPANISH STORIES/CUENTOS ESPAÑOLES: A Dual-Language Book, Angel Flores (ed.). Unique format offers 13 great stories in Spanish by Cervantes, Borges, others. Faithful English translations on facing pages. 352pp. 5⅜ x 8½. 0-486-25399-6

GARDEN CITY, LONG ISLAND, IN EARLY PHOTOGRAPHS, 1869–1919, Mildred H. Smith. Handsome treasury of 118 vintage pictures, accompanied by carefully researched captions, document the Garden City Hotel fire (1899), the Vanderbilt Cup Race (1908), the first airmail flight departing from the Nassau Boulevard Aerodrome (1911), and much more. 96pp. 8⅞ x 11¾. 0-486-40669-5

OLD QUEENS, N.Y., IN EARLY PHOTOGRAPHS, Vincent F. Seyfried and William Asadorian. Over 160 rare photographs of Maspeth, Jamaica, Jackson Heights, and other areas. Vintage views of DeWitt Clinton mansion, 1939 World's Fair and more. Captions. 192pp. 8⅞ x 11. 0-486-26358-4

CAPTURED BY THE INDIANS: 15 Firsthand Accounts, 1750-1870, Frederick Drimmer. Astounding true historical accounts of grisly torture, bloody conflicts, relentless pursuits, miraculous escapes and more, by people who lived to tell the tale. 384pp. 5⅜ x 8½. 0-486-24901-8

THE WORLD'S GREAT SPEECHES (Fourth Enlarged Edition), Lewis Copeland, Lawrence W. Lamm, and Stephen J. McKenna. Nearly 300 speeches provide public speakers with a wealth of updated quotes and inspiration–from Pericles' funeral oration and William Jennings Bryan's "Cross of Gold Speech" to Malcolm X's powerful words on the Black Revolution and Earl of Spenser's tribute to his sister, Diana, Princess of Wales. 944pp. 5⅜ x 8⅜. 0-486-40903-1

THE BOOK OF THE SWORD, Sir Richard F. Burton. Great Victorian scholar/adventurer's eloquent, erudite history of the "queen of weapons"–from prehistory to early Roman Empire. Evolution and development of early swords, variations (sabre, broadsword, cutlass, scimitar, etc.), much more. 336pp. 6⅛ x 9¼. 0-486-25434-8

AUTOBIOGRAPHY: The Story of My Experiments with Truth, Mohandas K. Gandhi. Boyhood, legal studies, purification, the growth of the Satyagraha (nonviolent protest) movement. Critical, inspiring work of the man responsible for the freedom of India. 480pp. 5⅜ x 8½. (Available in U.S. only.) 0-486-24593-4

CELTIC MYTHS AND LEGENDS, T. W. Rolleston. Masterful retelling of Irish and Welsh stories and tales. Cuchulain, King Arthur, Deirdre, the Grail, many more. First paperback edition. 58 full-page illustrations. 512pp. 5⅜ x 8½. 0-486-26507-2

THE PRINCIPLES OF PSYCHOLOGY, William James. Famous long course complete, unabridged. Stream of thought, time perception, memory, experimental methods; great work decades ahead of its time. 94 figures. 1,391pp. 5⅜ x 8½. 2-vol. set. Vol. I: 0-486-20381-6 Vol. II: 0-486-20382-4

THE WORLD AS WILL AND REPRESENTATION, Arthur Schopenhauer. Definitive English translation of Schopenhauer's life work, correcting more than 1,000 errors, omissions in earlier translations. Translated by E. F. J. Payne. Total of 1,269pp. 5⅜ x 8½. 2-vol. set. Vol. 1: 0-486-21761-2 Vol. 2: 0-486-21762-0

LIGHT AND SHADE: A Classic Approach to Three-Dimensional Drawing, Mrs. Mary P. Merrifield. Handy reference clearly demonstrates principles of light and shade by revealing effects of common daylight, sunshine, and candle or artificial light on geometrical solids. 13 plates. 64pp. 5⅜ x 8½. 0-486-44143-1

ASTROLOGY AND ASTRONOMY: A Pictorial Archive of Signs and Symbols, Ernst and Johanna Lehner. Treasure trove of stories, lore, and myth, accompanied by more than 300 rare illustrations of planets, the Milky Way, signs of the zodiac, comets, meteors, and other astronomical phenomena. 192pp. 8⅜ x 11. 0-486-43981-X

JEWELRY MAKING: Techniques for Metal, Tim McCreight. Easy-to-follow instructions and carefully executed illustrations describe tools and techniques, use of gems and enamels, wire inlay, casting, and other topics. 72 line illustrations and diagrams. 176pp. 8¼ x 10⅞. 0-486-44043-5

MAKING BIRDHOUSES: Easy and Advanced Projects, Gladstone Califf. Easy-to-follow instructions include diagrams for everything from a one-room house for bluebirds to a forty-two-room structure for purple martins. 56 plates; 4 figures. 80pp. 8¾ x 6⅜. 0-486-44183-0

LITTLE BOOK OF LOG CABINS: How to Build and Furnish Them, William S. Wicks. Handy how-to manual, with instructions and illustrations for building cabins in the Adirondack style, fireplaces, stairways, furniture, beamed ceilings, and more. 102 line drawings. 96pp. 8¾ x 6⅜. 0-486-44259-4

THE SEASONS OF AMERICA PAST, Eric Sloane. From "sugaring time" and strawberry picking to Indian summer and fall harvest, a whole year's activities described in charming prose and enhanced with 79 of the author's own illustrations. 160pp. 8¼ x 11. 0-486-44220-9

THE METROPOLIS OF TOMORROW, Hugh Ferriss. Generous, prophetic vision of the metropolis of the future, as perceived in 1929. Powerful illustrations of towering structures, wide avenues, and rooftop parks–all features in many of today's modern cities. 59 illustrations. 144pp. 8¼ x 11. 0-486-43727-2

THE PATH TO ROME, Hilaire Belloc. This 1902 memoir abounds in lively vignettes from a vanished time, recounting a pilgrimage on foot across the Alps and Apennines in order to "see all Europe which the Christian Faith has saved." 77 of the author's original line drawings complement his sparkling prose. 272pp. 5⅜ x 8½. 0-486-44001-X

THE HISTORY OF RASSELAS: Prince of Abissinia, Samuel Johnson. Distinguished English writer attacks eighteenth-century optimism and man's unrealistic estimates of what life has to offer. 112pp. 5⅜ x 8½. 0-486-44094-X

A VOYAGE TO ARCTURUS, David Lindsay. A brilliant flight of pure fancy, where wild creatures crowd the fantastic landscape and demented torturers dominate victims with their bizarre mental powers. 272pp. 5⅜ x 8½. 0-486-44198-9

Paperbound unless otherwise indicated. Available at your book dealer, online at **www.doverpublications.com**, or by writing to Dept. GI, Dover Publications, Inc., 31 East 2nd Street, Mineola, NY 11501. For current price information or for free catalogs (please indicate field of interest), write to Dover Publications or log on to **www.doverpublications.com** and see every Dover book in print. Dover publishes more than 500 books each year on science, elementary and advanced mathematics, biology, music, art, literary history, social sciences, and other areas.